BETTING ON IDEAS

BETTING ON IDEAS

Wars, Inventions, Inflation

Reuven Brenner

The University of Chicago Press
Chicago and London

Reuven Brenner teaches at the Université de Montréal. His first book, *History—The Human Gamble*, is also published by the University of Chicago Press.

HN
8
.B687
1985

The University of Chicago Press, Chicago 60637
The University of Chicago Press, ltd., London

© 1985 by The University of Chicago
All rights reserved. Published 1985
Printed in the United States of America

94 93 92 91 90 89 88 87 86 85   54321

Library of Congress Cataloging-in-Publication Data

Brenner, Reuven.
  Betting on ideas.

  Bibliography: p.
  Includes index.
  1. Social History. 2. War. 3. Economic history.
4. Social prediction.   5. Risk-taking (Psychology)
6. Chance.   7. Power (Social sciences) I. Title.
HN8.B73 1985   302   85-8750
ISBN 0-226-07400-5

To our two small boys, Adi and David

*Be it that the shattered fragments of
my thoughts will help you in the future.
But remember that . . .*

Ideas have been the most dangerous forces in the history of mankind.

William O. Douglas

Contents

Preface

*This whole book is but a draft—nay, but the draft of a draft. Oh,
Time, Strength, Cash, and Patience!*
> Herman Melville, *Moby Dick*

This book presents additional implications and applications of the views
presented in *History—The Human Gamble* (Brenner 1983b). To understand
this book, however, it is not necessary to be acquainted with the earlier one.

Chapter 1, which presents a theory of war, examines two questions:
When is the outbreak of wars more likely? What can one say about the
probability for recovery of the winners and of the losers? Of course, the
theory is not deterministic: while it shows that the probability of war
increases when some nations outdo others, it also shows that chance deals
the decisive blow at this point.

In contrast to customary approaches, wars are not examined from an
"economic," "sociological," "psychological," or "political-science" point
of view. Rather, realizing that events do not happen in categories (categories are merely accidental human inventions) and that wars, like all acts,
begin in people's minds, I look at the circumstances in which people may
either gamble on wars or change their minds and gamble on peace. Thus
wars are examined here not by looking at "nations," but by looking at the
behavior of individuals. The question raised is this: When is it more likely
that many people (call them a "nation") will suddenly gamble on one idea
rather than another? What is it, in general, that triggers a "thought," a
change of mind?

The views presented here suggest that policies designed to maintain
people's status and the international status quo diminish the probability of
wars *even* if the distribution of wealth within a society and the distribution of
power among societies are unequal. Whether or not the design of policies
that can achieve these goals is always feasible is another question, and
answers to it are examined in the final, concluding chapter of the book.

The advantage of the approach presented in chapter 1 becomes clear in
the chapters that follow. They deal with a wide variety of acts and facts
related to gambling, crime, innovations in technology, arts and the law,
productivity and inflation. Yet this wide range of evidence is examined by
using the same approach that is used to shed light on wars, and its seeming
applicability increases our confidence in the validity of the model. Briefly,
the advantage of my views is that in contrast to those of other social scientists

which have been developed to deal with wars and nothing else, mine make a wide variety of predictions which, though totally unrelated to wars, can be easily verified.

Only simple, everyday words are used in this book—for several reasons. First, as Theodore Roosevelt, speaking as president of the American Historical Association in 1912, said, "writings are useless unless they are read, and they cannot be read unless they are readable." Second, I really do not see the point of using technical words when dealing with everyday aspects of human behavior. Yet, at times (mainly in the appendixes), some statements are translated to mathematics. The main reason for the translation is to allow the reader, when I deal with complex subjects like wars, thinking, productivity and others, to perceive immediately and precisely where I have oversimplified, made unwarranted assumptions, or omitted motives that should have been included. Readers would have much more difficulty finding out what is hiding behind the words if the translations to mathematics were absent. Third, I share the views of Alfred North Whitehead, who once wrote: "The aim of science is to seek the simplest explanation of complex facts. We are apt to fall into the error of thinking that the facts are simple because simplicity is the goal of our quest. The guiding motto in the life of every natural philosopher should be: 'Seek simplicity and distrust it.'"

Indeed, I think that the ideas presented in this book *are* simple and *should* be distrusted.

Acknowledgments

Many people have encouraged me to continue the line of argument presented in my first book and make an effort to apply it to new domains or look for additional evidence that could contradict it. This is what I try to do in this book. In chapter 2 I was aided by my wife Gabrielle; chapter 4 was written by her alone.

We owe thanks to many people both for their encouragement and their comments—we really needed them: William Baumol, Gary Becker, James Buchanan, Dennis Carlton, Léon Courville, Colin Day, Antal Deutsch, Leonard Dudley, Jean-Marie Dufour, Marc Gaudry, Charles Goodhart, Barbara Haskel, A. Heertje, Jack Hirshleifer, Elisabeth Johnson, Timur Kuran, Robert Lacroix, Robert Lafrance, A.F.K. Organski, William McNeill, Frederick Mishkin, Joel Mokyr, G.O.W. Mueller, Douglass C. North, Frederic Pryor, T. W. Schultz, and Rodrigue Tremblay. Special thanks are due to Wendy Dobson who taught me how to write.

We are also grateful to our research assistants, Richard Guay and Maxime Trottier, for their excellent and dedicated work, and to our secretaries, Nicole Laporte and Carmelle Paré, for their patience through many drafts.

We thank the Ecole des Hautes Etudes Commerciales and the University of Montreal, for continuous support. A grant from the Social Sciences and Humanities Research Council of Canada helped the final revisions of the manuscript done at the Centre de Recherche et Developpement en Economique.

1   Why Do Nations Engage in Wars?

The life of nations no less than that of men is lived largely in the imagination.

Enoch Powell, 1946

The consequences of wars can rarely, if ever, be predicted. Recognizing this fact, political scientists have built models to try to answer these questions: Why do wars break out? What circumstances are more likely to lead to their outbreak?

Although there is a vast literature devoted to war, one can find only a few basic models on which the analyses have been based. A brief survey of these models will serve as the point of departure for this chapter. Next it will be shown that these models are *not* significantly different but reflect separate aspects of a general view of human behavior, which is presented in the second section. In contrast to these models, this view of behavior does not start from the assumption that predictable behavior can be attributed to *nations*. Rather, it examines the motivations and identifies the circumstances under which many individuals (call them "a nation") become more likely to agree to go to war, start new alliances, or break up old ones. This view of human behavior suggests that in all circumstances there is an unpredictable element—the ideas of an individual (who becomes "the" political leader)—that will have an effect on whether or not the nation will gamble on peaceful or other solutions. The third section shows how this view of human nature can explain the "Davids and Goliaths" and the "phoenix factor" (the quick recovery of losers from the ashes). In section 4 a brief summary of evidence on these two phenomena and further opinions about wars are presented and the question of how the incidence of wars can be diminished is examined. In the concluding section the methodology for examining the implications of the model is discussed.

1. Three Models for Analyzing the Outbreak of Wars

Political scientists have offered three basic models to explain why nations decide to fight: "balance of power," "collective security," and "power transition." They are briefly summarized below.[1]

Balance of Power

The balance-of-power model suggests that there will be peace when power is approximately equally distributed among members of major alliances. And the contrary: the probability that there will be an outbreak of war is greater when there are significant differences in the distribution of power. According to this view of the world, the relatively powerful country, whose power is increasing, will attack the weaker adversaries. Organski (1968) summarizes the mechanism of this model:

> Given large numbers of nations with varying amounts of power each one striving to maximize its own power, there is a tendency of the entire system to be in balance. That is to say the various nations group themselves together in such a way that no single nation or group of nations is strong enough to overwhelm the others, for its power is balanced by that of some opposing group. As long as the balance can be maintained, there is peace and the independence of small nations is maintained. [p. 274]

The implicit assumption behind this mechanism is that nations are motivated by their desire to maximize their power. The recognition of every nation that this is the goal of every other nation leads the ones endowed with less resources to enter into alliances in order to give a signal to the relatively stronger to avoid starting a war. The balance of power is thus maintained by the strategy of forming and canceling alliances.

There have been several criticisms of this model; Organski and Kugler (1980) summarize them:

> If the equilibrium is disturbed, the system favors adjustments that will return it to equilibrium. But if this jockeying process . . . cannot reallocate the power loads sufficiently to obtain a roughly equal distribution among the major actors in the system, then one nation possessing decisive strength and *uninvolved* in existing coalitions, will step in on the weaker side and redress the balance, thus rendering the system ultrastable . . . Those who espouse the balance-of-power model do not clearly explain why one nation should be exempt from the otherwise universal rule of wanting to take advantage of its superiority to expand its power at the expense of others . . . [and] Do all nations really wish to maximize their power? One cannot help noticing variations, over time, in the degree to which they have wished to do so. [pp. 16–17; italics in original]

The views presented in the second section will explain both these variations *and* the fact that one nation, the most powerful, while motivated by the same universal rule as the other nations, can be expected to act differently.

Collectively Security

According to the "collective security" model the world is divided into aggressors and peace-loving nations. The latter, if they form a coalition, are assumed to be stronger than the aggressors. However, if some of these nations defect because of either greediness or fear, the chance of war increases.

This model makes the following assumptions: that aside from the aggressor all other nations love peace; that there is an agreement on who the aggressor is; that, as in the previous model, changes in alliances have an effect on the probability of war; and that collective security provides "security" rather than "peace." These assumptions have been criticized on the grounds that there rarely seems to be an argument on who "the aggressors" are (the conflicts in the Falklands and the Middle East are recent examples), and that the assumption that one nation is the constant aggressor is arbitrary.

The second section explains how the goal of "security" is compatible with the goal of "maximization," and why aggressors may turn into peace-loving nations, and the contrary.

Power Transition

According to the "power-transition" model, developed in the fifties, peace is viewed as being preserved even when there is an unequal distribution of power between nations. Aggression is likely to occur when some nations suddenly outdo others. The mechanism of this model is summarized by Organski (1968):

> At the very apex of the pyramid is the most powerful nation in the world. . . . Just below the apex of the pyramid are the great powers. The difference between them and the dominant nation is to be found not only in their different abilities to influence the behavior of others, but also in the differential benefits they receive from the international order to which they belong. . . . the powerful and dissatisfied nations are usually those that have grown to full power after the existing international order was fully established and the benefits already allocated. These parvenus had no share in the creation of the international order, and the dominant nation and its supporters are not usually willing to grant the newcomers more than a small part of the advantages they receive . . . The challengers, for their part, are seeking to establish a new place for themselves in international society, a place to which they feel their increasing power entitles them. . . . [pp. 364–67][2]

According to this model the source of war is to be found in the unexpected different rates of growth of the members of the international system.[3] The rate of growth must be significant because it is assumed that if development is slow, the problem that arises when one nation catches up with the dominant one may have a greater chance of being resolved. The arguments presented in section 2 suggest that the goal of maximization leads the ones who are perceived to be becoming relatively weaker to gamble on aggressive startegies and those who are perceived to be becoming relatively stronger to look for security, and that indeed only significant, sudden changes seem to matter.

Comments

While the models presented above differ, they also share a number of features. Each of them seems to ascribe *predictable* behavior to "nations." In the balance-of-power model, the leaders of a nation seek to maximize power, while in the collective-security model, they try to prevent aggression, and in both cases it seems to be implicitly assumed that the population shares its leaders' goals. The third model, too, makes a prediction as to the behavior of nations, namely, that when there is a significant change in the power structure, a "nation" may gamble on an aggressive strategy.

But several questions come to mind: "nations" are an abstraction. Since "nations" are composed of individuals, what assumptions about individual behavior could lead to these predictions? Why would one leader act in one way and another in a different way? The model presented in the next section attempts to answer these questions. It shows that by starting from a simple set of assumptions about the behavior of *every* individual, one can understand how people make decisions and why they change their minds, on political strategies in particular. Since the decision either to start or to try to prevent wars is a manifestation of their ways of thinking when facing risks, the basic question one must raise is how people form their opinions in such circumstances. The next section examines this fundamental question.

2. How Do People Make Up Their Minds?

Thought, whither dost thou lead me? For it is a universally admitted truth that it is unhealthy to think and that true wisdom lies in not thinking at all.

Anatole France; *The Revolt of the Angels*

"Thinking," defined as "bets on ideas," is the subject that I tried to shed light on in my previous book, and the one that I also try to shed light on here, though I am examining new directions. Summarized briefly (a more formal statement of these views appears in the Appendixes to this chapter), the views suggest that bets on new ideas are triggered when customary ways of behavior fail to produce expected results and lead to the perception of a loss in one's relative standing in society.[4] One can attribute this adaptive behavior to "envy," "ambition," or "fear of being hindered by others" (recall Mandeville's "Envy itself and Vanity Were Ministers of Industry"). This perception and these sentiments lead to the following types of behavior:

— people may start to participate in games of chance that they have previously shunned;
— they may commit a crime, or an act not in accordance with existing customs;
— they may gamble on new (i.e., noncustomary) ideas in business, science,

technology, the arts, and politics. The appeal of gambling on political ideas—going to war in particular—is greater when a whole group's standing in a society or the society of nations has suddenly been significantly worsened. This prediction will be explored in this chapter.[5]

And the contrary: when aspirations are more than fulfilled, and people suddenly outdo their fellows, they
— will tend to take out insurance that they previously shunned;
— may avoid committing a crime that they contemplated before;
— may avoid betting on new ideas.

A few more observations can also be made: When aspirations are realized, the relatively rich spend a relatively greater fraction of their wealth on insurance, while the relatively poor spend a relatively greater fraction on games of chance in which they may lose relatively small sums, but have a chance to win big ones. These predictions are made by assuming that individuals try to do their best when facing uncertain prospects, and that the distribution of wealth is pyramidal—that is, there is a small "upper class," a larger "upper-middle class," a still larger "lower-middle class," and so forth. Implicitly, however, additional assumptions are made: the incentive to gamble on new ideas appears when an individual's position in the distribution of wealth has suddenly been (or is expected to be) significantly worsened. The incentives disappear if either customs or redistributive taxes lead to expectations that the individual will be compensated for his loss.[6]

There are some good reasons for such customs or such a system of taxation to evolve (eventually), if these views of human behavior are accurate. For they predict, as we have observed, that when an individual's situation is relatively worsened, he may gamble not only on an entrepreneurial act but also on a criminal one, which is costly for a society. Moreover, if a whole group's economic position has been significantly diminished, the probability increases that its members will gamble on political—even revolutionary—ideas advocating a redistribution of wealth in their favor. Since such gambles, too, are costly for the society, policies or customs requiring redistribution of wealth may provide remedies that maintain the stability of the society, though simultaneously diminishing its creativity. But notice that if in such societies one happened to have a "fluctuating" life (in spite of all insurances), and eventually becomes creative, one may be perceived in a negative light and may be discouraged from implementing one's ideas. For a rise in the distribution of wealth is expected to motivate those who fell behind to gamble both on additional entrepreneurial acts and on criminal ones. Both are viewed as costly since they are expected to lead to further fluctuations.[7]

As the evidence presented in *History—The Human Gamble* and here suggests, historical events (some of them due to chance) seem to determine why members of one society gamble on one system (of maintaining a somehow achieved order), and why others, *for a while*, choose the alternative (of tolerating some disorder). The evidence also suggests that unless fluctuations occur that are beyond human control, mechanisms exist which eventually lead the second types of societies, too, to gamble on institutions that will maintain their stability. However, as long as these two types of societies coexist, one can predict that the ones where there are fewer mechanisms for preventing fluctuations in people's position in the distribution of wealth will be the ones where inventions will be made and implemented. The nature of some of these inventions will be examined in this book.

Political Ideas and Foreign Affairs

War is not an instinct but an invention.

José Ortega y Gasset, 1930

Until now, for simplicity's sake, the predictions made were based on the assumption that only changes in the internal distribution of wealth have mattered. However, similar predictions can be obtained if instead of assuming that the internal distribution of wealth has changed and the external stayed constant, one assumes that the internal has stayed constant and the external one has changed. Only the words used to describe the resulting events differ.[8]

Suppose that there has been a change in the distribution of wealth among some countries. For example, the U.N. might have voted for changing the borders between two countries, or one country has occupied another one, or one country has suddenly become stronger than its neighbors (because somebody there invented a more sophisticated weapon). What will people do in the country whose relative position in the distribution of wealth has suddenly diminished?

According to the arguments presented in the previous section, it is likely that an individual—call him a political leader—will come up with some ideas that constitute a challenge to the allocation of wealth and advocate a redistribution. These are ideas that the rest of the population now has a *greater* incentive to gamble on: they represent the fixed point of reference in the midst of the sudden chaos. The idea may take many forms: advocating peaceful negotiations for compensations, justifying terrorist acts against the country that became richer, advocating new alliances as threats against the now richer country, or advocating a strategy of war. The model does not permit a prediction as to the type of strategy chosen: it only identifies the circumstances that are more likely to lead to a gamble on any of these strategies, including the strategy of war. (Notice that, in terms of the model, the idea of war is always, latently, lurking in people's minds. However, an act of war is initiated when the fear of being outdone by "fellows," an idea that previously had only crossed people's minds, has become a reality.) These circumstances seem to be the same as those postulated by the "power-transition" model: when sudden, significant changes in the distribution of wealth among countries occur, the probability of an outbreak of war increases, because many people in the country whose position in the distribution of wealth has worsened (or is expected to worsen), are likelier to change their minds.

These arguments and the views presented in the previous section also show how a "nation" can exhibit a predictable behavior: when its expectations are frustrated, "the right man at the right time" may come up with a political slogan that the rest of the population may gamble on—it is in this sense that one can speak about a "nation's" behavior. Paraphrasing Orwell's (1945) more poetic language on the behavior of nationalists, one could say that the political leader secures more power and more prestige not for himself, but for the nation in which he has chosen to sink his own individuality (remember that Orwell defines a nationalist leader as one who thinks solely, or mainly, in terms of competitive prestige [p. 157]).[9]

A nation's behavior can also be predicted when its wealth suddenly increases (because of a discovery of gold or oil, for example). This nation now has greater incentives to insure itself, rather than to gamble. The political ideas that may become popular in such countries are to provide help to other countries or to serve as "policemen" in some regions and thus try to restore stability; this is exactly how wars may be prevented. The differences in the behavior of the richer country are not simply a matter of differences in "tastes." Rather, understanding the incentives that motivate people (even if the incentives have not been articulated), may help the people in the richer country to realize that they now run a greater risk of being attacked. In order to diminish this risk, those who have become richer may either spend more on weapons, or transfer wealth to the poorer country, and thus try to maintain order.[10]

These arguments show that the "collective-security" model, too, is thus consistent with the implications of the views presented here. The different attitudes toward wars stem from the different and changing position of some nations in the international distribution of wealth.

3. On Davids and Goliaths; or, What Is "Wealth" and What Is "Power"?

The story of David and Goliath is well-known, and history books are filled with exciting tales of small armies winning against great powers. If the view of human behavior spelled out here is accurate, the phenomenon is not as strange as it may seem at first sight.

While the outcomes of wars cannot be predicted from the views presented here, what can be predicted is that people who feel suddenly threatened have greater incentives to become entrepreneurs and gamble on new ideas—on a surprising gadget in David's case (if one wants to take the story literally, rather than as a good allegory of human behavior). At the same time these views also predict that Goliaths (i.e., those perceived as stronger) will have lesser incentives to gamble on new ideas, and greater incentives to insure themselves.

So what is "power" and what is "wealth"? The somewhat surprising, and for some perhaps uncomfortable, answer is that we really do not and cannot know precisely what these terms mean. In the three models used in political science (described in the first section), only the term "power" is used. Yet the term is never defined precisely. When attempts have been made to examine the three theories rigorously (in Organski and Kugler 1980, for example), the measures that have been used in order to translate the abstract notion of "power" to a concrete notion of everyday life have been either a country's population size and its GNP or the ability to raise taxes, measures that in standard economic analysis are used as indicators of "wealth."[11] Yet in estimating a nation's ability to win wars, this concept of wealth is not very useful as a measure of "power." The view of human behavior presented here shows why: a nation that gambles on an individual's political ideas, and thus stands behind him, may be "wealthier" than another in which the population is split in its political beliefs. People belonging to the first nation will make greater efforts, since they bet more frequently on innovative ideas.[12] But only a fraction of the second nation follows this behavior pattern. Thus the first nation may turn out to be more powerful in times of war even if both its population and its GNP are smaller than that of the other country, and ex post measures of GNP and population size might thus turn out to be inaccurate indicators of the probability of victory. The important concept, if the arguments in the preceding sections are valid, is *not* the static notion of wealth, but rather the dynamic one. Indeed, this observation has been made by some social scientists (although in a different language): Wright (1942) in his massive study of war agrees with Charles E. Merriam on the following point:

> The power does not lie in the guns, or the ships, or the walls of stone, or the lines of steel. Important as these are, the real political power lies in a definite common pattern of impulse. [Notice what Merriam calls "real".] If the soldiers choose to disobey or even shoot their officers, if the guns are turned against the government, if the citizenry connives at disobedience of the law, and makes of it even a virtue, then authority is impotent and may drag its bearer down to doom. [p. 117]

and Bateson (1972) writes:

> Since both Americans and English respond most energetically to symmetrical stimuli, we shall be very unwise if we soft-pedal the disasters of wars. If our enemies defeat us at any point, that fact ought to be used to the maximum as a challenge and a spur to further effort. [p. 104]

Not that the observation on the importance of morale is novel: the "crumbling walls of Jericho" gives us a lesson on its importance, albeit in an allegorical language. The radios and other channels of communication in today's cold and hot wars fulfil the same role—of trying to destroy morale—as the shouting through trumpets did in biblical times. In his classic *On War* Clausewitz makes similar points and constantly criticizes the view of those who try to describe war tactics and how to win wars in terms of wealth and "base," the latter being comprised of the number of soldiers, their equipment, communication with the home country, etc. While he admits that the idea of a "base" is a necessity for strategy, he concludes that such a belief in the decisive effect of the form of attack is contrary to common sense, since it leaves out human behavior and directs "the attention only upon material forces, while the whole military action is penetrated throughout by intelligent forces and their effects" (1832, p. 184). Moreover, he emphasizes that "if we desire to defeat the army, we must proportion our efforts to his powers of resistance. This is expressed by the product of two factors which cannot be separated, namely, *the sum of available means and the strength of the Will*" (p. 104, italics in original). "The inequality between physical forces [of two armies] might be such that it could be balanced by the moral forces" (p. 124)—in a later section more will be written on Clausewitz's views.

Concerning the notions of "power" and "wealth" one may thus raise this question: although perceptions of changes in the distribution of wealth provide the motivation for changes in behavior, it has been argued here that "wealth" cannot be measured. How, then, can my views serve as a positive framework for the analysis of wars? The answer is this: since the only things that one can ever examine are the effects of significant *changes* in people's relative wealth, one might make comparisons between two situations and make predictions even if wealth is not measured properly.

Internal Turmoil and External Adventure

One of the most frequent charges that can be encountered in history textbooks is that internal turmoil may lead political leaders to gamble on a strategy of war. It is a straightforward exercise to show that indeed the view of human nature suggested here does allow one to make such a prediction.

When we use the term "internal turmoil" we are saying that more criminal acts and more frequent bets on political ideas, revolutionary in particular, are taking place. Such acts occur with greater frequency when significant changes in people's position in the distribution of wealth occur, and they decrease the stability of a society. Suppose that a significant group in the population have become "déclassé" and thus are more likely to gamble on a political idea that leads to expectations for redistributing wealth in their favor. Since this group's wealth has also diminished relative to that of a neighboring country, one political idea that this group may gamble on is that of a strategy of war against the neighboring country.

Various writers have believed in the validity of the view that internal turmoil increases the probability of wars, although they sometimes state it differently and do not seem to have a precise view of human nature on which to rely in order to support their beliefs. Clausewitz (1832), for example, wrote that "national hatred, which is seldom wanting in our wars, is a substitute for personal hostility in the breast of individual opposed to individual" (p. 186). Schumpeter, in *Imperialism and Social Classes*, also notes that wars may break out because of domestic political difficulties. He illustrates his views by discussing the situation in ancient Rome, where "an unstable social structure . . . created a general disposition to watch for pretexts for war . . . and to turn to questions of foreign policy whenever the discussion of social problems grew too troublesome for comfort. The ruling class was always inclined to declare that the country was in danger, when it was really only class interests that were threatened" (1919, p. 69).[13] But Schumpeter considers this case, of internal turmoil leading to wars, as being fundamentally different from another in which wars break out because of competition for resources. The views presented here show the uniformity below the surface: people are more likely to gamble on the idea of war when they perceive sudden changes in the distribution of wealth, national or international.[14]

What Happens after the War?

There seem to be several opinions as to what might happen in countries after fighting a war.[15] Keynes (1919) was the proponent of one view. He maintained that in some circumstances the gap that developed between winners and losers as a result of war would increase, and losers would fall further and further behind. Keynes's views are discussed in detail later in this section. A second view, that of Norman Angell (1933), suggested that as a result of war all nations lose, but winners lose less. The gap between winners and losers, then, stays stable and may last for long periods of time. T. W. Schultz (1961) offered another opinion. While most economists were pessimistic about the recovery of Western Europe after World War II, Schultz was optimistic. He argued that while much physical capital was destroyed, "human capital" (i.e., the knowledge embodied in human beings) was not, and thus recovery was likely to be swift. However, Schultz did not offer a hypothesis that could have explained the differential rate of recovery of winners and losers whose populations had similar levels of education (a point recently raised by Mancur Olson 1980, 1982). Organski's views (summarized in Organski and Kugler 1980) were different: while he asserted that losers suffer more than winners, and are thus in a worse position immediately after conflicts, in the long run they catch up with winners, since they recover more quickly:

> Most unexpected and interesting is the discovery that, after wars, the active losers catch up with winners in comparatively short order, and

that the system of international power begins to behave as one would have anticipated had no war occurred. We cannot explain the phenomenon; we do not know why losers rise from the ashes as they appear to do. [p. 142]

This unexplained phenomenon was known as "the phoenix factor."

The views presented here suggest a straightforward explanation for this phenomenon and make additional predictions. Suppose that while both losers and winners have lost part of their wealth, the losers have lost more. They are thus more likely to continue to make greater efforts and engage more frequently in entrepreneurial acts (relative to the people whose position in the distribution of wealth has unexpectedly improved) until their wealth is restored to its customary, expected level. One would expect a faster recovery in countries that lost wars relative to others that won them and may have stayed closer to their customary position in the international distribution of wealth.

Without providing a rigorous view of human creativity to justify his arguments, Olson (1982) explained the differential recovery of European countries after World War II in the following way:

. . . stable societies with unchanged boundaries will accumulate more organizations and collusions for collective action over time. The reason is that, as time passes, more of those groups that can organize will have enjoyed the fortunate circumstances and able political entrepreneurship that is needed for organization, whereas the interest of organizational leaders in maintaining their position insures that organizations with selective incentives will not disappear unless destroyed by upheaval or war . . . [these] distributional coalitions slow down a society's capacity to adopt new technologies and establish barriers to entry that reduce a society's capacity to reallocate resources quickly in response to changing conditions, and thereby reduce the rate of economic growth. [pp. 144–45]

These observations seem consistent both with the previous arguments and the other implications of my views. It has been predicted that when wealth diminishes and the distribution of wealth changes, people become more likely either to gamble on individual effort to restore their wealth, or, if they happen to be part of a group, may engage in political pressures to redistribute wealth in their favor. Since such interest groups are in general destroyed in countries that have lost a war, as Olson suggests, it becomes more likely that after wars individuals will bet on *individual* effort to restore their wealth. In contrast, in the winning countries the interest groups may not have been destroyed, and if a fraction of the population lost its wealth, it still may use traditional political methods to try and redistribute wealth in its favor. Thus one would expect a quicker recovery in the losing rather than in the winning country. The quicker recovery will continue until individuals in the losing country begin to gamble on new political ideas around which new interest groups can be organized that will try to maintain their group's position in the wealth-distribution stable.[16]

Yet it is important to point out that losers may not *always* recover more quickly: the conditions a peace treaty imposes on the losers may play a central role in the process of recovery. Let us discuss this conclusion by illustrating it with Keynes's pessimistic views of the Versailles treaty after World War I (as described in *The Economic Consequences of the Peace*).

Keynes (1919) points out that "the Carthaginian peace is not *practically* right or possible" (p. 23; italics in original), since the treaty went against the spirit, purpose, and intention of President Wilson's fourteen points. While in one of the addresses before the Congress (on 11 February) it was explicitly stated that

> There shall be no annexations, *no contributions, no punitive damages* . . . Self-determination is not a mere phrase. It is an imperative principle of action which statesmen will henceforth ignore at their peril . . . Every territorial settlement involved in this war must be made in the interest and for the benefit of the populations concerned, and not as a part of any mere adjustment or compromise of claims amongst rival States. [quoted in Keynes 1919, p. 39; italics in original]

the draft of the treaty ignored these implicit promises, and Germany was required to pay far greater sums than those its leaders could expect from President Wilson's guidelines. Keynes makes several calculations showing that Germany does not have the ability to pay the sums according to the conditions of the treaty and emphasizes:

> I cannot leave this subject as though its just treatment wholly depended either on our own pledges or on economic facts. The policy of reducing Germany to servitude for a generation, of degrading the lives of millions of human beings, and of depriving a whole nation of happiness should be abhorrent and detestable—abhorrent and detestable, even if it were possible, even if it enriched ourselves, even if it did not sow the decay of the whole civilised life of Europe. Some preach it in the name of justice. In the great events of man's history, in the unwinding of the complex fates of nations, justice is not so simple. And if it were, nations are not authorised, by religion or by natural morals, to visit on the children of their enemies the misdoings of parents or of rulers. [p. 142]

Let us forget the moral appeal of these sentences and consider the practical implications of the harsh peace treaty in terms of my view of human nature. Its conditions imply not only that Germany's wealth has unexpectedly diminished (since the conditions of the treaty were inconsistent with previously made promises), but also that incentives to make greater efforts have been nonexistent. For anything that the older and younger German generations may have produced beyond some minimum level of subsistence would have been taken away from them and redistributed to the Allies for years. But the necessary condition for recovery is that people must expect that the fruits of their lucky hits will belong to them. If they cannot hold such expectations, the incentives to gamble on new ideas in business, or at work, do not exist. Instead, incentives exist to gamble either on illegal acts (nondeclaration of incomes in particular), or on new political ideas. The reason for the increased incentives to do the latter is that harm done by political mistakes of the past can be remedied by new political gambles—in particular, by disobeying a treaty.

The similarity between Keynes's summary in the chapter "Europe After the Treaty" and the implications of the model thus seem obvious:

> This chapter must be one of pessimism. The treaty includes no provisions for the economic rehabilitation of Europe—nothing to make the defeated Central empires into good neighbours, nothing to stabilise the new states of Europe, nothing to reclaim Russia; nor does it promote in any way a compact of economic solidarity amongst the Allies themselves; no arrangement was reached at Paris for restoring the disordered finances of France and Italy, or to adjust the systems of the Old World and the New . . . The danger confronting us, therefore, is the rapid depression of the standard of life of the European populations . . . [But] men will not always die quietly. For starvation, which brings to some lethargy and a helpless despair, drives other temperaments to the nervous instability of hysteria and to a mad despair. [pp. 143–44]

Keynes wrote these words in 1920.

So winners may lose, and losers may win: "winning" and "losing" seem, at times and in a historical perspective, illusions for a fleeting moment (and World War II shows how fleeting such moments may be)—just as the myth of the death and the rebirth of the phoenix suggests.[17]

4. How Have Others Viewed the "Causes" of Wars? Learning from Machiavelli's *Prince*

The idea that changes in relative positions in the international distribution of wealth might be the cause of wars is not novel. In his *History of the Peloponnesian War* Thucydides implied that such was the case when he wrote that the growth of Athenian power terrified Sparta and forced it into war:

> The Athenians made their Empire more and more strong . . . [until] finally the point was reached when Athenian strength attained a peak plain for all to see and the Athenians began to encroach upon Sparta's allies. It was at this point that Sparta felt the position to be no longer tolerable and decided by starting the present war to employ all her energies in attacking and if possible destroying the power of Athens. [p. 77]

He thus explained the policy of Tissaphernes, the Persian king, who tried to achieve balance between the two powers.[18] In a recently published book, Michael Howard (1983) reaches the conclusion that the model implicit in Thucydides' views is still accurate; only the names of the actors have changed.

Later Polybius, in the *Histories*, relies on a similar argument to explain Hiero's policies. First, Hiero allied himself with the Romans against Carthage. But a few years later he became worried about the Romans' success and sent assistance to Carthage. Polybius explains:

> . . . it was in his own interest for securing both his Sicilian dominions and his friendship with the Romans, that Carthage should be preserved, and that the stronger power should not be able to attain its ultimate object entirely without effort. [quoted in Waltz 1954, p. 199]

Continuing this brief chronological survey of opinions on the causes of wars let us turn to Machiavelli's *Prince* and point out the strong similarities between his and my views on human nature, wars, and "ideal" governments. The reason for pointing out similarities and not differences is that I could not find any of the latter.[19]

First, Machiavelli explains why it is more difficult to rule in new principalities than in hereditary ones: " . . . because it is enough merely not to neglect the institutions founded by one's ancestors and then to adapt policy to events . . . the natural prince has less reason and less need to give offence . . . And in the antiquity and persistence of his rule memories of innovations and the reasons for them disappear" (pp. 33–34). In contrast, in new principalities difficulties arise: "What happens is that men willingly change their ruler, expecting to fare better. This expectation induces them to take up arms against him; but they only deceive themselves, and they learn from experience that they have made matters worse" (pp. 34–35).

But if the people in countries acquired by force share the same language and have similar customs as the intruders, it is easy to hold them since "so long as their old ways of life are undisturbed and there is no divergence in customs, men live quietly; as we have seen in the case of Burgundy, Brittany, Gascony and Normandy, which have been with France for so long . . . " (p. 36). The Prince will be hated if "he is rapacious and aggressive with regard to the property and the women of his subjects. He must refrain from these. As long as he does not rob the great majority of their property or their honour, they remain content" (p. 102).

How can these statements be translated to the views presented in the previous sections? If in hereditary principalities the distribution of wealth is kept stable by customs, laws, and police, the probability that innovations in the organization of social institutions will gain currency is relatively small. The cooperation of occupied people is explained by this same insight: Machiavelli suggests that leaving these people more or less "undisturbed" diminishes the probability of upheavals. Since customs, laws, and police protect people's economic position, incentives to gamble on new ideas, political in particular, are absent. In contrast, new principalities arise because of struggles among groups which are started by those whose expectations have been disappointed. When a principality was occupied, "This is what happens: as soon as a powerful foreigner invades a country all the weaker powers give him their support, moved by envy of the power which has so far dominated them" (p. 38).

How can conquered provinces be kept quiet? Machiavelli discusses two methods: either through troops and punishments, or through settlements. He considers the latter preferable for two reasons: "Settlements do not cost much . . . [and the Prince] injures only those from whom he takes land and houses to give to the new inhabitants, and these victims form a tiny minority, and can never do any harm since they remain poor and scattered" (p. 37). In contrast, if the Prince sends troops, expenses are far higher and the army's presence is a source of constant friction. While today one may disagree with these views, the disagreement may stem not from Machiavelli's views of human nature, but from the changed costs and benefits of the two methods. First, settlements today may cost a fortune (Israel's experience is an example); second, since the population is much larger, the victims of evacuation may form more than a tiny minority, and, while they may remain poor, they may not remain scattered but may get organized and commit criminal or terrorist acts. But the changes in the relative advantage of the two methods then have to do with increased numbers, and not with the basic motives underlying human behavior.

Machiavelli emphasizes the roles of individuals and of groups, and the interaction among them, as well as the role of chance, when he discusses the probability of making innovations in the organization of institutions:

> The very fact that from being a private citizen he has become a prince presupposes either ability or good fortune . . . And when we come to examine their actions and lives, they do not seem to have had from fortune anything other than opportunity. Fortune, as it were, provided the matter but they give it its form . . . Thus for the Israelites to be ready to follow Moses, in order to escape from servitude, it was necessary for him to find them, in Egypt, enslaved and oppressed by the Egyptians. For Romulus to become king of Rome and founder of his new country, he had to have left Alba and been exposed to die when he was born. Cyprus needed to find the Persians rebellious against the empire of the Medes, and the Medes grown soft and effeminate through the long years of peace. Theseus could not have demonstrated his prowess had he not found the Athenians dispersed. The opportunities given them enabled these men to succeed, and their own exceptional prowess enabled them to seize their opportunities; in consequence their countries were ennobled and enjoyed great prosperity. [pp. 50–51]

The similarity between these arguments and examples and those stressed in the previous sections regarding the role of individual and social change are evident. This view of human behavior led Machiavelli to conclude that in Italy " . . . so many things conspire to favour a new prince, that I cannot imagine there ever was a time more suitable than the present . . . in order to discover the worth of an Italian spirit, Italy had to be brought to her present extremity. She had to be more enslaved than the Hebrews, more oppressed than the Persians, more widely scattered than the Athenians; leaderless, lawless, crushed, despoiled, torn, overrun; she had to have endured every kind of desolation" (p. 134). Why? Since in these circumstances, Machiavelli responds, the probability of honoring a new prince is greater.

Machiavelli also notes that unless some significant changes occur, innovators are unlikely to succeed: "It should be borne in mind that there is nothing more difficult to handle, more doubtful of success, and more dangerous to carry through than initiating changes in a State's constitution. The innovator makes enemies of all those who prospered under the old order, and only lukewarm support is forthcoming from those who would prosper under the new. Their support is lukewarm partly from fear of their adversaries, who have the existing laws on their side, and partly because men are generally incredulous, never really trusting new things unless they have tested them by experience" (p. 51).

What role does chance play when one looks at the lives of individuals? Machiavelli writes: "It can be observed that men use various methods in pursuing their own personal objectives, that is glory and riches. One man proceeds with circumspection [he insures himself?], another impetuously [he gambles?]; one uses violence [he commits criminal acts?], another stratagem [he gambles on new ideas in business, science or politics?] . . . and yet everyone, for all this diversity of method, can reach his objective. It can also be observed that with two circumspect men, one will achieve his end, the other not; and likewise two men succeed equally well with different methods, one of them being circumspect and the other impetuous" (p. 131). According to Machiavelli, whether or not one succeeds depends on whether or not one correctly identifies the circumstances in which one lives, and adapts himself: . . . "if time and circumstances change, he will be ruined because he does not change his policy . . . If he changed his character according to the time and circumstances, then his fortune would not change" (p. 132).

However, Machiavelli concludes that this is easier said than done. For two reasons: neither "do we find any man shrewd enough to know how to adapt his policy," nor can any man do so, because "men are obstinate." Machiavelli was right: when the circumstances suddenly change, no one can form a clear idea about one's interests; one just gambles. As to being "obstinate," I showed in my previous book that the persistence of customs and people's rigid behavior can be attributed to the fact that when no significant alterations in the distribution of wealth occur, people will tend to imitate their ancestors' behavior and copy their way of life, rather than deviate from established ways. This seemed to have been Machiavelli's view, too: "Men nearly always follow the tracks made by others and proceed in their affair by imitation, even though they cannot entirely keep to the tracks of others or emulate the prowess of their models. So a prudent man must always follow in the footsteps of great men and imitate those who have been outstanding" (p. 49).

Machiavelli's views of the cyclical history of nations (i.e., the descent from prosperity to adversity and back) stem from his views of human nature and seem close, too, to the implications that can be derived from mine. It has been argued that when one nation suddenly falls behind another, its people become more likely to gamble on all forms of entrepreneurial acts, and the ascent may start. And the contrary, if fortune suddenly smiles on a nation, and that nation becomes suddenly richer (say, because gold or oil was discovered), its people have fewer incentives to gamble and greater incentives to just insure themselves and imitate others. The slow descent may start when circumstances again remain stable, and the altered behavior becomes habitual. Machiavelli's evaluation of the probabilities for survival of newly acquired principalities is related to this view. He points out that when new principalities are acquired by arms and prowess, the probability of success in forming a stable principality is greater than if the new principality was acquired just by chance, or by the new prince insuring himself by using armies of other principalities.

> Private citizens who become princes purely by good fortune do so with little exertion on their own part; but subsequently they maintain their position only by considerable exertion . . . Such rulers rely on the goodwill and fortune of those who have elevated them, and both these are capricious, unstable things. They do not know how to maintain their position, and they cannot do so. They do not know how, because . . . private citizens are incapable of commanding; they cannot because they do not have loyal and devoted troops of their own. Then again, governments set up overnight, like everything in nature whose growth is forced, lack strong roots and ramifications. So they are destroyed in the first bad spell. [pp. 53–54]

Machiavelli illustrates his view with an interpretation of the first part of the story of David and Goliath:

> I would also like to recall to mind an allegory from the Old Testament . . . David offered Saul to go and fight Goliath, the Philistine champion, and Saul, to inspire him with courage, gave him his own weapons and armour. Having tried these on, David rejected them, saying that he would be unable to fight well with them and therefore he wanted to face the enemy with his sling and his knife. In short, armour belonging to someone else either drops off you or weighs you down or is too tight. [p. 85]

The lesson is clear: at times, when facing suddenly altered circumstances, people should gamble on new ideas, rather than rely on customary behavior (represented in the allegory by Saul's military strategy).

A note on the morality of Machiavelli's and my arguments. There is no morality in them. When circumstances are stable, people fight among themselves within limits defined by customs and laws. But when disturbances occur in people's position in the distribution of wealth the probability that people will abandon the established ways and resort to force increases. Here is how Machiavelli summarizes his views on the choice between the two methods: " . . . there are two ways of fighting: by law or by force. The first way is natural to men, and the second to beasts. But as the first way often proves inadequate one must needs have recourse to the second. So a prince must understand how to make a nice use of the beast and the man. The ancient writers taught princes about this by an allegory, when they described how Achilles and many other princes of the ancient world were sent to brought up by Chiron, the centaur, so that he might train them his way. All the allegory means, in making the teacher half beast and half man, is that a prince must know how to act according to the nature of both, and that he cannot survive otherwise" (p. 99). If my views are correct, this is true not only for princes, but for all men. For people are guided by customs when circumstances are stable (and Machiavelli considers these guidances as being "natural to men") and resort to force when customs suddenly break down (and Machiavelli considers this behavior as "natural to beasts").

Why did people feel so uneasy about Machiavelli's views? One reason may be that the implication of his views of human nature is that there is no obvious form of "ideal" government that one can recommend. Notice that his and my views suggest not only that the price a society pays for increased creativity is to tolerate some disorder (expressed in increased crime rates, increased probability of social unrest and of wars—there is no manna from heaven in this model), but also that once some order is achieved and people get accustomed to it, there is no way one can say that one form of order is inferior to another.[20] The ideas that people have gambled on in order to achieve that order (ideas that later generations call "customs")[21] are the sole regulators of morals and define the words "good" and "bad."[22] Not surprisingly, as one of the best observers of human behavior, Shakespeare had similar insights. The ideals of the Tudor statesmen are expressed in these lines from *Troilus and Cressida*:[23]

> O! when degree is shaked
> Which is the ladder to all high designs,
> The enterprise is sick. How could communities,
> Degrees in schools, and brotherhoods in cities,
> Peaceful commerce from dividable shores,
> The primogenitive and due of birth,
> Prerogative of age, crowns, sceptres, laurels,

But by degree, stand in authentic place?
Take but degree away, untune that string,
And, hark!   What discord follows;

Clausewitz's philosophy of war had a profound influence on European military and political thought in the nineteenth century, and it is considered to rank with the works of Francis Bacon, Machiavelli, Hobbes, Hume, Adam Smith, and Marx. According to Clausewitz decisions to wage war and to conclude peace are made by sovereigns. Though their interests and their ambitions matter, he never mentions that their acts may come in response to sudden internal or external turmoil, or that wars may be rooted in the suddenly crumbling—national or international—orders (although he argues that any theory of war must take into account the interaction among "people," "leaders," and "the army," and he constantly emphasizes the human element as a decisive factor in war). Thus, viewed in the light of my views, Clausewitz's philosophy reflects only part of the picture, namely, the crucial roles individuals, leaders in particular, and chance play before and during wars.[24]

Schumpeter's (1919) views are in sharp contrast to Clausewitz's. Instead of leaders and chance he emphasizes classes (making reference to pyramidal class structures) and customs:

> The explanation [for wars] lies . . . in the vital needs of situations that molded peoples and classes into warriors—if they wanted to avoid extinction—and in the fact that psychological dispositions and social structures acquired in the dim past in such situations, once firmly established, tend to maintain themselves and to continue in effect long after they have lost their meaning and their life-preserving function. . . . Imperialism is thus atavistic in character. It falls into that large group of surviving features from earlier ages that play such an important part in every concrete social situation . . . It is an atavism in the social structure, in individual, psychological habits of emotional reaction. [pp. 84–85]

Or, as Schumpeter puts it elsewhere in the text, when the fixed order of earlier times is disturbed, wars may break out.[25] The similarities between these views and mine are clear: according to both it is a sudden change from a somehow established order that leads people to gamble on strategies of war, and the existence of a power or class structure is necessary. The differences are clear, too: Schumpeter not only lacks a precise model of human behavior in whose light he could discuss his views, but he also seems to leave no roles in his argument for individuals and for chance. Moreover, he seems to suggest that the impulse to go to war rests on "primitive contingencies of physical combat" that will gradually disappear, "washed away by the new exigencies of daily life." This view is in sharp contrast with mine in which the "impulse" to gamble on the strategy of war is inherent in the human condition, since whenever leapfrogging occurs the probability of an outbreak of wars increases.

In their recent book Organski and Kugler (1980) make a first attempt to analyze the "causes" of wars by statistical methods. Their data refer to four periods: the Franco-Prussian war (1870), the Russo-Japanese war (1904), World War I (1913), and World War II (1939). With this limited data set, they analyze which of the three models ("balance of power," "collective security," or "power transition") seems most consistent with the evidence. Their result is unambiguous:

> The mechanisms that make for major wars can be simply summed up. The fundamental problem that sets the whole system sliding almost irretrievably toward war is the differences in rates of growth among the great powers and, of particular importance, the differences in rates between the dominant nation and the challengers that permit the latter to overtake the former in power. It is this leapfrogging that destabilizes the system. [p. 61]

Michael Howard's (1983) view is similar. According to him the causes of World War I are in essence no more complex or profound than those of any previous European war, or indeed than those described by Thucydides as underlying the Peloponnesian war. He writes:

> In Central Europe, there was the German fear that the disintegration of the Habsburg Empire would result in an enormous enhancement of Russian power—power already becoming formidable as French-financed industries and railways put Russian manpower at the service of her military machine. In Western Europe there was the traditional British fear that Germany might establish a hegemony over Europe which, even more than that of Napoleon, would place at risk the security of Britain and her own possessions; a fear fuelled by the knowledge that there was within Germany a widespread determination to achieve a world status comparable with her latent power. Considerations of this kind had caused wars in Europe often enough before. Was there . . . anything different about 1914? [p. 10]

Howard concludes that many of the German people, and in 1939 nearly all the British, felt justified in going to war, not over any specific issue, but to restore their power, while in recent times the Soviet Union is aspiring to maintain its status rather than aiming at world hegemony. Whether or not the last perception is accurate is another matter, but there is little doubt that in Howard's views, too, the leapfrogging process seems to play a central role. Earlier, Halford Mackinder (1962) held similar views: "The great wars of history—we have had a world war about every hundred years for the last four centuries—are the outcome, direct or indirect, of the unequal growth of nations" (pp. 1–2). But if no such leapfrogging occurs, that is, when the distribution of wealth is *stable* (even if unequal), one cannot make any prediction. Indeed this was Organski and Kugler's (1980) second and final conclusion:

> Wars seem to occur both when adversaries are equal and unequal in power. In this initial step, then, power distributions are obviously not a predictor of the coming war. The introduction of the concept of one nation surpassing another in power as the independent variable [in explaining the probability of the outbreak of hostilities] brings us an important new piece of information. [p. 49]

And what is a better illustration of the accuracy of this observation than Morison's (1965) view of the American Revolution:

> Make no mistake; the American Revolution was not fought to *obtain* freedom, but to *preserve* the liberties that Americans already had as colonials. Independence was no conscious goal, secretly nurtured in cellar or jungle by bearded conspirators, but a reluctant last resort, to preserve "life, liberty and the pursuit of happiness." [1:248]

War: An Invention

In her essay, "Warfare Is Only an Invention—Not a Biological Necessity," Margaret Mead wrote that warfare, like dueling, is just an invention by which societies permit their young men either to accumulate prestige or avenge their honor. "Honor" and "prestige" are words that indicate one's relative position in the distribution of wealth; the word "invention" refers to an *idea*. The fact that customs permit duels to take place (in some societies) in which one can "avenge one's honor" suggests that the honor was previously unexpectedly blemished. Since honor is part of one's reputation, and reputation is part of one's wealth, this observation, too, is consistent with the implications of the model, since it implies that a gamble on warfare is permitted when part of one's wealth is lost.

Mead is not the only anthropologist who wrote about wars. Wright (1964) notes that there is a school of anthropology, represented by W. H. Rivers and G. Elliot Smith (and elaborated in the works of W. J. Perry), that contends that war was invented in predynastic Egypt, along with agriculture, social classes, and human sacrifice. While there is no evidence at all to support this view, it may be useful to note that one can plausibly link the widespread use of agricultural methods, social classes, human sacrifice, and wars. For, as argued and demonstrated by many social scientists (Boserup 1965, 1981; Cohen 1977; and others summarized in Brenner 1983), the widespread use of agricultural methods and the invention of new ones can be linked to sudden population growth (as can the idea of "human sacrifice" as a way of possibly controlling this growth). Sudden population growth, by decreasing the individual share of wealth and altering its distribution, is linked in turn to changed incentives and thus to inventions in general.[26] Thus there might be a correlation among all the phenomena mentioned by Wright, although their root cause may be a sudden growth in population.[27]

What alternative ideas can people gamble on—instead of starting a war—when the distribution of wealth (or the status quo) is altered? The possibilities are numerous: "universal brotherhood," for one. This idea, however, requires trust among nations that may take too long to build.[28] Compensating the nation that has fallen behind may be another idea, and as the example below suggests, such a plan can work and can maintain stability for long periods of time. Herskovits (1940) describes the following customs among tribes in the Nilgiri hills of India:

> The members of this tribe were musicians and artists for the three neighboring folk of their area, the pastoral Toda, the jungle-dwelling Kurumba, and the agricultural Badaga. Each tribe had clearly defined and ritually regulated obligations and prerogatives with respect to all the others. The Toda provided the Kota with ghee for certain cere-monies and with buffaloes for sacrifices at their funerals. The Kota furnished the Toda with the pots and knives they needed in their everyday life and made the music essential to Toda ceremonies. The Kota provided the Badaga with similar goods and services, receiving grain in return. They stood in the same exchange relationship with the forest Kurumba, but these latter, who could only provide meagre material compensation . . were able to afford the Kota supernatural protection, since the Kurumba were dreaded sorcerers, so feared that every Kota family must have their own Kurumba protector against the magic which others of this tribe might work against them. [p. 157]

What was the cause of the Kota's belief? How could it emerge and become a custom?

The emergence of this belief can be explained in the following way. Since the Kurumba were, or became, relatively poor, they were the ones more likely to start a war. Realizing the threat, the Kota, in an attempt to diminish this probability, gambled on the idea of transferring payments to them. The Kurumba's threat (which meant diminished wealth for the Kota) might have been the source of the custom and the cause of the emergence of the belief that the Kurumba were "dreaded sorcerers," an idea that both the Kota and the Kurumba had the incentives to gamble on. Indeed, as the epigraph of this chapter states, the life of nations is lived largely in the imagination.[29]

This example is disquieting. For very long periods of time, the tribes mentioned above maintained their stability by compensating potential "aggressors," and they did not gamble on the idea of teaching the aggressors that gambling on another set of ideas might be more beneficial for all parties concerned. So can one be hopeful that nations today will be different from these "primitive" tribes? Can the current international situation be stabi-lized by continuously subsidizing poorly endowed nations and thereby maintaining some already achieved order? The concluding chapter discusses this issue.

What Is the Evidence on the "Phoenix Effect"?

The observation that winners turn into losers has been made by many historians. Here is how Cipolla (1976) summarizes the decline of Spain:

> The decline of Spain in the seventeenth century is not difficult to understand . . . Spain, as a whole . . . became considerably richer . . . during the sixteenth century . . . The riches of the Americas provided Spain with purchasing power but ultimately they stimulated the development of Holland, England, France, and other European countries . . . At the end of the sixteenth century, Spain was much richer than a century earlier, but she was not more developed—"like an heir endowed by the accident of an eccentric will." . . . In the meanwhile . . . a century of artificial prosperity had induced many to abandon the land, schools had multiplied, but they have served mostly to produce a half-educated intellectual proletariat who scorned productive industry and manual labor and found positions in the bloated state bureaucracy which served above all to disguise unemployment. Spain in the seventeenth century lacked entrepreneurs and artisans . . . [1976, pp. 233–36]

The similarity between this description and my predictions about human nature is rather startling (although not really unexpected; Hegel noted long ago the phenomenon of the helplessness of the victors). In a recent article titled "The Smallest Nation Has a Rare Problem: Too Much Wealth," Gigot (1983) describes how the Nauru phosphate income ($25,000 for each of the 5,000 Nauruan citizens) fell like manna from heaven bringing much laziness and waste. Nauru imports all its workers: Filipinos build houses, Australians pilot their jets (their airline loses $30 million per year), Gilbert Islanders dig phosphate and catch fish, later selling them to Nauruans waiting in cars. Most Nauruans work as bureaucrats or not at all. Gigot also describes one of their surviving customs, called *bubuji*; when good fortune smiles on a Nauruan, friends *ransack* his property. Both the laziness and this custom can be easily explained in light of my views. The first phenomenon is expected to occur when people unexpectedly become significantly richer. The second phenomen represents one strategy by which the stability of a small, isolated community can be maintained. In its absence, a sudden increase in wealth would increase the incentives of those who fell behind to gamble on criminal acts. The custom of *bubuji*, permitting the redistribution of wealth, seems to maintain the society's stability even today. But once the $25,000 income started pouring in there were other changes. The Nauruans not only stopped working, but, imitating other people they also started to import expensive new food, which, since they are unaccustomed to it, is making them sick and obese. (But apparently they do not care; see appendix 1.1 for an explanation of this apathy.) No wonder that one Nauruan stated to Gigot that to him the phosphate, instead of suggesting wealth, suggested a graveyard.

In contrast to such declines in entrepreneurial activities, Toynbee (1966) describes the rise of entrepreneurial activity in Persia:

> . . . the classical Persian poetry had been written in the course of the half-millenium between the break-up of the Abbassid empire and the political reunification of Iran in the Safavi Empire. During this period, . . . in spite of the consequent insecurity of life and destruction of wealth, a fractured Iran, like a fractured Greece and Italy, excelled in the arts. [p. 93]

Notice, however, that according to my hypothesis it is not "in spite of the destruction of wealth," but rather *because* of it that arts have excelled. McNeill (1963) wrote:

> The first notable tremor in the balance between China's commercial and landed interests occurred after the Sung emperors lost north China to Jurchen invaders (1127). Thrown back upon . . . resources of the south, China in the later Sung period saw a notable development of riverine and maritime trade. Great cities arose on the south China coast and along the Yangtze; and growing numbers of merchant vessels set sail for southeast Asia and the Indian Ocean. [p. 525]

As to more systematic evidence, Organski and Kugler (1980) have also examined the effects of wars, and as mentioned earlier, their "most unexpected and interesting . . . discovery [is] that, after wars, the active losers catch up with winners in comparatively short order" (p. 142). Moreover, they argue that aids to the losers had *negative* rather than positive effects on the rate of recovery:

> Had there been a direct relationship between aid and recovery, and if one controlled for population, growth rates would show increases as a result of aid. Had aid intensity been a factor one would also expect that growth rates would show strong gains after those years when recipients received particularly large gifts . . . [But] such relationship as may exist is negative: the countries that received most of the aid for the longest period performed worst. The United Kingdom received much more aid, on a total and per capita basis, than France; France received much more than Italy, Italy much more than Germany, and Germany much more than Japan. Yet it was Japan that enjoyed the most rapid rate of recovery, followed by Germany, Italy and France, with the United Kingdom bringing up the rear. It is, therefore, very hard to credit the conviction that foreign assistance and recovery are closely associated. [pp. 143–44]

Thus the recovery from the ashes cannot be attributed to aid but to an alternative mechanism. The views presented in this chapter suggest such a mechanism.[30]

5. Can the Incidence of Wars Be Diminished?

Let us turn to the practical question: Does the viewpoint presented here suggest a strategy for eliminating wars, or at least diminishing their incidence? The answer seems simple (although as shown in the concluding chapter of the book, implementing it may be difficult): leaders of governments should try to avoid changes in people's position in the distribution of wealth, *both* internal *and* external. This statement does not imply that *if* the distribution of wealth has been changed for reasons beyond human control, political leaders should not *start* initiatives—they should. The statement implies only that there should be an institution in existence to check abuses of power. The difference between the two cases, when institutional changes are warranted and when they are not, lies in the perception of the situation: does the struggle between the powers simply reflect an attempt to maintain a somehow achieved status quo, or does it reflect a fundamental instability in the international system? The answer to this question is given in the concluding chapter, after the role demographic changes play in the analysis is clarified. Before reaching that stage it may be useful to contrast my answer with answers that others have given to the question of how to prevent wars.

Somewhat superficially, one can classify these writers in two categories: some who saw that "the root of all evil is man" and looked for the causes of war in the behavior of individuals, separated from the times they lived in, and others who looked for the causes of war within the structure of states and within the state system. Augustine, Spinoza, Reinhold Niebuhr, and Morgenthau seem to belong to the first category, since they start political analysis by looking at individual behavior, while Rousseau, Bernard, and many sociologists (for a summary of their views see Waltz 1958 and Wright 1964) belong to the second since they seemed to suggest that human behavior can be better understood by studying society, or the society of states. The model presented here combines the two views in a straightforward way and suggests that the behavior of man cannot be analyzed without his history. Human behavior cannot be understood unless one looks simultaneously at the individual *and* at the society he lives in.

In spite of the fact that the writers mentioned above have only looked at one side of the coin, there are similarities between their views and mine. Augustine had observed the importance of "self-preservation" in the hierarchy of human motivations. So had Spinoza: to him the end of every act is the self-preservation of the actor, and acts that do seem altruistic are not. Regard for others is due to the recognition that mutual assistance is necessary to one's preservation. According to Augustine bad things, wars in particular, happen because human reason and will are "defective," while according to Spinoza they happen because of our dual nature: the conflict between "reason" and "passion." Niebuhr and Morgenthau seem simply to suggest that man is defective and that the origin of wars lies in the "dark, unconscious sources in the human psyche" (using Niebuhr's words in *Beyond Tragedy*).

These are, of course, words and opinions and *not* facts. What does it mean to say that we are "defective," or that our "reason" fights with our "passions"? Would a man be a "man" if he were not "defective," or if he had only "reason" or only "passions"?

In contrast to views based on these abstract, undefined words, this chapter suggests a way of defining human behavior without attaching any normative connotation to it. Yes, the model does imply that people may go to wars and commit crimes. And yes, these acts can be defined as matters of "passion," if passion is defined as the reaction to unexpected changes in one's status in society. But no, no distinction can be made between "reason" and "passion," if by "reason" one means all the beneficial inventions, discoveries, and innovations that people have made (as Spinoza suggested). Since it was the reaction to unexpected changes in their status that led people to make them, "reason" and "passions" thus become synonyms.

According to Rousseau man's behavior, which the aforementioned writers have taken as cause, is in great part determined by the society in which he lives, and "society" is defined as a political organization. Thus the major causes of war should stem from the state system itself. Bernard (1944), a sociologist, wrote that one needs to know what dangerous social conditions actually to correct in order to prevent wars. Mark May (1943) wrote that in order to have peace one must learn loyalty to a larger group, and before one can learn loyalty, the "thing" to which one should be loyal must be invented. (This is a rather abstract suggestion. How will people know that what one human mind will invent will be "the" thing to which they now should be loyal? Remember that Hitler did offer such a "thing" and that some rabbis said long ago that "when the shepherd goes astray, the flock errs after him.")

Durbin and Bowlby's (1939) views seem closest to the model presented here since they have argued that war is due to the expression in and through group life of the transformed aggressiveness of individuals, that personal character derives from environment as well as from inherited nature, and that it may be possible to change the character of adult behavior by changing the environment in which one unchanged hereditary element develops. According to my views the change in people's perception of their position in the distribution of wealth leads them to become more willing to gamble on new ideas, on going to war in particular (call it "transformed aggressiveness"?). The goal of self-preservation is our "unchanged hereditary element," and behavior *can* be changed if customs or taxes exist which keep the distribution of wealth stable, thus preventing disturbance in some already achieved order. (Remember the old saying on being slaves to chance, kings, and desperate men?)[31]

Conclusions

Two clear-cut implications can be derived from the arguments presented here with respect to wars. First, that the probability of a war increases when one nation "overtakes" another: such shifts have destabilizing effects. Second, there is a mechanism that induces the losers to recover quickly, and this does *not* happen because the winners help them; such help may in fact only slow down the rate of recovery. These two predictions seem to be supported by both the statistical and the nonstatistical evidence that is available.

The model also makes clear the role of individuals, politicians in particular, in situations in which the probability of war increases. Recall that this probability increases because people became more likely to bet on ideas that could help them restore their relative position in the national or international distribution of wealth. New ideas will be offered by many people, and a priori one cannot tell which ideas the population will find most appealing, those that suggest wars or others that suggest peace. The individual on whose ideas the population gambles plays a central, random, role in this process.

This conclusion is in contrast with that of Organski and Kugler (1980) who wrote that policy makers "along with the rest of us, are merely actors, speaking their lines on cue without being able to change the script" (p. 221). Yet this conclusion is based on neither their analyses nor their findings: in their empirical analyses they have only analyzed situations in which wars *did* break out. Thus they were able to learn something about the *relative* relevance of the three traditional models that explained wars, but *not* about the role of individuals. In order to learn something about their role, Organski and Kugler also had to look at the number of times in which one nation "passed" another yet *no* war occurred. Instead peaceful solutions seemed to be found. The explanation for this difference stems exactly from the role played by the individual on whose ideas the population gambles.[32]

Finally, some methodological points. Wars have not been examined in this chapter from an "economic," "sociological," "psychological," or "political science" point of view. Realizing, rather, that for men of "flesh and bone" wars begin in their *minds* (as *all* acts do), I have examined war by looking at the circumstances in which people may either gamble on a strategy of war, or *change their minds* and gamble on that of peace. This methodology, of avoiding the currently popular classifications and trying to suggest a uniform approach to examine human behavior, is not as unusual as it may seem at first sight. Marc Bloch (1953), in his unfinished notes, suggested more than forty years ago that "As for *homo religiosus, homo oeconomicus, homo politicus*, and all that rigmarole of Latinized men, the list of which we could string out indefinitely, there is grave danger of mistaking them for something else than they really are: phantoms which are convenient provided they do not become nuisances. The man of flesh and bone, reuniting them all simultaneously, is the only real being" (p. 151).[33]

As Organski and Kugler's (1980) study makes clear, there are not much data available to test various hypotheses concerning wars. Thus it may be difficult to distinguish between various theories that can explain only wars and nothing else. The view presented here has the relative advantage of looking at human behavior in terms of some fundamental characteristics and making numerous predictions that are related neither to wars nor to specific circumstances and can thus be (and have been) more easily examined. Were one to find that a wide range of both contemporary and historical evidence does not contradict the predictions of the model, one could use it with greater confidence; the rest of the chapters in the book concentrate on presenting and discussing the evidence.

Last but not least, recall what Carl Friedrich wrote in *Inevitable Peace*: "It is not usually recognized by people who discourse upon war and peace that any general theory of war implies a general view of history. Nor have they always been aware of the fact that you cannot usefully discuss the problems of how to maintain peace if you have no theory of war" (p. 54). Whether or not one agrees with my views is one question, but my writings seem at least to pass this test. In *History—The Human Gamble*, one can find out what my view of "history" is, a view that can also be discovered as my story unfolds in the next chapters.

Appendix 1.1:
On Probability, Thinking, and Stress

Mr. Jones
> *how can you know what here goes on*
> *behind this flesh-bright frontal bone?*
> *here are the world and god, become*
> *for all their depth, a simple* Sum.

The Mirror
> *well, keep the change, then, Mr. Jones,*
> *and, if you can, keep brains and bones,*
> *but, as for me I'd rather be*
> *unconscious, except when I see*
>> Conrad Aiken, *The Coming Forth by Day of Osiris Jones*

Let us clarify the meaning of two words used in the model in this chapter as well as in my previous book: "probability" and "thinking."

The reason for clarifying the latter word is mainly this: readers may ask how such a simple model can provide insights into so many aspects of human behavior. My answer is simple. Since this model suggests how people think, and since thinking is at the root of what our behavior is all about, one should be able (if my model is accurate) to shed light on anything human beings have done in the past. If some readers are worried about the model's simplicity they should remember Ockham's suggestion that the unnecessary multiplication of assumptions should be avoided and look for the simplest model that sheds light on the widest range of evidence—a methodology I have tried to stick to.

1. On Probability

The terms "gamble" and "probability" (denoted in the next appendix's mathematical expressions by p) have been used in two totally different contexts in this chapter. When speaking about lotteries and insurance, the word "probability" has been used to represent the notion of a probability distribution—the assignment of probabilities to a set of related events, events that could be repeated many times. These probability distributions were assumed to be the same for every individual who got involved in these staged events.

But the term "probability" has also been used in a totally different context: to represent the degree of belief an individual has attached to the success or the failure of implementing an idea.

How this probability "got there" in our minds, I do not know. Our ancestors have said that it was due either to "divine inspiration" or to "Satan's temptation." The words I have used in the text *seem* clearer: namely, "lucky hit" or "the dark sides of our psyches." However, these are just words that we are *accustomed* to; they are neither more precise than the ones used by our ancestors, nor do they reveal more about the nature of the acts themselves. Indeed "divine inspiration" and "lucky hits" seem to be synonyms.

Thus one could raise the question that if this is the case, in fact, I did not define what probability *is*, so how could I build theories of gambling, insurance, and creativity without such a definition? The question is legitimate and my answer pragmatic.

Just like mathematicians, who have written so many books on probability without defining either randomness or probability (but instead have given an operational definition, namely, "the definition of randomness and probability is that which obeys the theories they derive from it"), I have used a similar approach. The mathematical theory of probability begins after probabilities have been assigned to elementary events, and my theory also starts at that point. How probability is assigned is not discussed either by the mathematicians or by me because that would require an intrinsic definition of the randomness of events, which is impossible (see last chapter). However, Laplace and other mathematicians have made the point that although individual random events were meaningless, the *distribution* of these events was not and could be subject to "probability theory." It is in this sense that the term "probability" has been used in the discussion of participation in lotteries.

In the discussion of creativity and crime, "probability" represents a degree of belief that stayed the *same* in the two situations that have been compared. Only one's position in the distribution of wealth has changed. This led to the conclusion that people "changed their minds." But if I wanted to be more precise I should have written that "the mind" (i.e., the idea) was "there" (in "the back of one's mind"); it was merely "latent." The philosophical implication of this statement, clarifying in what sense we are "preprogrammed," I put aside.

Before proceeding it may be interesting and useful to note that the customs of many societies indicate a relationship among fighting, games of chance, and beliefs (religious in particular), and that the word "play" had its origin in the spheres of ethics, law, and religion. For example, Huizinga (1944) points out that the word "play," the German *"pflegen,"* and the Dutch *"plegen"* are derived from a combination of the Old English *"plegan,"* and the Old Frisian *"plega,"* and that their meaning was "to vouch or stand guarantee for, to take a risk, to expose oneself to danger" (p. 39). He discusses many other examples as well, in Hebrew and in Greek. He also notices that "with many peoples dice-playing forms part of their religious practices . . . In the *Mahābhārata* the world itself is conceived as a game of dice which Siva plays with his Queen . . . The main action of the *Mahābhār-ata* hinges on the game of dice which King Yudhistira plays with the Kauravas . . . [and] a whole chapter [of it is devoted] to the erection of the dicing-hall—sabhā" (p. 57). Indeed Huizinga notes that in archaic language Divine Will, destiny, and chance were equivalent concepts, and "Fate" may be known by eliciting some pronouncements from it. An oracular decision of this kind is arrived at by trying out the uncertain prospects of success. You draw sticks, or cast stones, or prick between the pages of the Holy Book, and the oracle will respond. In Exodus 28:30, Moses is bidden "to put in the breastplate of judgement the Urim and Thummim [nobody knows what they were, although Huizinga mentions that the word "urim" has affinities with a root that means casting lots, shooting as well as justice, law (yore, thorah)] . . . Likewise in Samuel 14:42, Saul orders lots to be cast as between himself and his son Jonathan . . . Pre-Islamic Arabia also knew this kind of sortilège. Finally, is not the sacred balance in which Zeus, in the Iliad, weighs men's chances of death before the battle begins, much the same? . . . This weighing or pondering of Zeus is at the same time his judging . . . " (p. 79). In Greek iconography "Diké (justice) frequently blends with the figure of Nemesis (vengeance) just as she does with Tyché (fortune)" (p. 94).

Harkabi (1983) makes a similar observation when he writes that "the recurrent blessing, 'God will help,' is not necessarily a prayer beseeching God to gird His loins and come to the warriors' aid; rather it is a wish for propitious circumstances. This factor . . . is said to influence the outcomes of battles and the course of history; all the great commanders of history prayed for good fortune likewise" (p. 41). So maybe it is true that our motivations are like those of our ancestors; only the vocabulary we use, borrowed from scientific achievements, is different. Moreover, it may also be that many people in the past have perceived human behavior in exactly the same way as I do; only the language they used to describe it was different and later generations misinterpreted it. Indeed, many writers have pointed out that the main difficulty in interpreting ancient events is that to the great despair of historians men fail to change their vocabulary every time they change their customs.

2. On Thinking

Do our thoughts and emotions dictate our behavior, merely using the brain as a servant to implement our aims? As suggested by both the arguments and the evidence presented here and in my previous book, the answer seems positive (making one reservation for the case in which, due to overbearing circumstances, one's aims become incomprehensible [see Brenner 1983b, chap. 1]).

How the human mind works has been examined by both psychologists and neurobiologists from various viewpoints. For a long time psychologists have followed the approach of not asking what is going on in people's minds, arguing that it is impossible to check whether or not the answers are accurate. The question they raised was the following: How do people react to various changes in circumstances? This question is similar to the one I have raised, but here the similarity between my methodology and those of the psychologists ends. My approach to making predictions on random events, which shows what is going on in people's minds when circumstances suddenly change, is simple and precise. However, in contrast to psychologists, I do not look at people's behavior in laboratories, since according to my views such evidence cannot provide reliable information.

The approach can also deal with many elements that psychologists have left out of their analysis, and by using ordinary, everyday language (avoiding words like "cognition," "conation," "affect," and using instead "thinking," "wanting," "emotions"). Moreover, it can deal with what Hadley Cantril (as quoted by Taylor 1979) admitted psychologists could not: " . . . I am sadly aware of my own inability to capture even remotely the wonder that mind is, especially when I think of such products of mind as the Fifth Symphony, the Sermon on the Mount, *The Brothers Karamazov* or the calculus. Whatever it is that enables mind to create and to appreciate such marvels seems to elude almost completely the crude nets of any psychological jargon." The approach presented here shows that chance and suffering produce these "products." The opinions of some psychologists should be noted: Taylor (1979) attributes to Theodore Schneirla of the American Museum of Natural History the view that all behavior can be traced to two responses: approach and withdrawal. According to Schneirla, one constantly weighs possibilities and considers losses and gains. The model reflects such behavior and suggests in what circumstances these trade-offs are solved (see appendix 1.2).

Neuroscientists and chemists have also expressed their views on the human mind. If my views are correct, then the neuroscientists who stated that brain research favors determinism are wrong (for a summary of their views see Taylor 1979). Chance and necessity mold into one in the human brain, and no logical distinction can be drawn between them. Several neuroscientists and chemists have expressed this view. "The enormous complexity . . . of the interactions possible . . . suggests that . . . output may not be predictable on the basis of unit properties alone" (Professor F. O. Schmitt, head of MIT's Neuroscience Research Program, as quoted in Taylor 1979, p. 65); "[thought is] 'active uncertainty.' It is a casting about among alternatives and an assigning of probabilities to different outcomes which is the constitutive feature of thought" (John Dewey, as quoted in Taylor 1979, p. 273).

It also follows that the argument of brain chemists that some substances regulate and create emotion must be carefully interpreted. If one finds that some level of chemical substance in the body is associated with depression, and then administers a drug to change that level, can one reach the conclusion that the "cause" of the behavior has been touched? The answer is negative, since one should not confuse the cause of a disease with its treatment. If one is depressed because one's expectations were not fulfilled, does a drug that corrects the resulting chemical inbalance touch "the" cause of the behavior? (See the next section for a discussion on why the chemical inbalance may be a result of stress.) The logic of such a conclusion reminds one of a Russian story (told by Professor R. Lewontin in a *New York Review of Books* article) about the psychologist who proves that fleas hear with their legs by training them to jump on command and then observes that they no longer respond when their legs are amputated.

However, some implications of my model seem consistent with the views of some biologists less than a century ago, Hans Driesch among them, who believed that life and living processes can never be explained by the laws of physics and chemistry. Dreisch postulated a "vital force" to explain the ability of the living to form and repair themselves (see Taylor 1979, p. 297).

3. On Memory, Stress, and Thinking

According to the views presented here, people think (that is, they gamble on new ideas) when their expectations are not realized.

But what happens when people live and expect to live in stable circumstances? I do not deal with this issue in this book, but I have dealt with it extensively in the previous one, where it has been shown that when people expect to live in stable circumstances, they are left only with the incentive to memorize the behavior of their ancestors and imitate it. The evidence supporting these views has also been presented there. Thus, according to my views, a sharp distinction can be made between the incentives to "memorize" (which exist when circumstances are expected to stay stable) and the incentives to "think" (which emerge when circumstances suddenly, significantly change). Of course, in everyday life, we perform both acts simultaneously since we expect some circumstances to stay relatively stable, but others change suddenly.

The distinction between "thinking" and "memorizing" has been made by social scientists, although instead of using the term "memorizing" they have spoken of the formation of habits. Bateson (1972) suggests that

> Samuel Butler was perhaps first to point out that that which we know best is that of which we are least conscious, i.e. that the process of habit

formation is a sinking of knowledge down to less conscious and more archaic levels. The unconscious contains not only the painful matters which consciousness prefers to not inspect, but also many matters which are so familiar that we do not need to inspect them. Habit, therefore, is a major economy of conscious thought. We can do things without consciously thinking about them . . . No organism can afford to be conscious of matters with which it could deal at unconscious levels. This is the economy achieved by habit formation. [pp. 141–43]

Bateson illustrates these arguments by the following, well-known, example: "I was watching a . . . carpenter-architect at work . . . I commented on the sureness and accuracy of each step. He said, Oh, that. That's only like using a typewriter. You have to be able to do that without thinking" (p. 148).

Where in our bodies our memories are stored, I do not know. Recent work suggests the existence of both "muscle" and "skin" memory, leading to the conclusion that the entire body remembers, "mind" being thus a property of the entire body rather than of just the brain (see Taylor 1979, pp. 141–43).

What happens when suddenly, due to changed circumstances, we become worse off? My model suggests that we then gamble on a new idea. But "who" exactly in our body is this "we," and what the new idea is that it gambles on—I cannot say. The ideas examined in this book and in my previous one refer to the creative act, which may be either positive or negative.

However, psychologists and physicists have also emphasized that stress can lead to illness, too: that busy executives are liable to stomach ulcers and high blood pressure seems to be a well-known fact; and some physicists believe that the same stress will evoke one type of illness in one person, and another in a second person. That skin diseases have an "emotional origin" also seems to be generally recognized. Taylor (1979) also points out that there is evidence that childhood stress either increases "emotionality" or crushes it, and that for adults there is a relationship between stress and all forms of illnesses. He quotes Thomas H. Holmes who built a scale for measuring the number of changes in a person's life (as an indicator for stress) and found that the 10 percent who had experienced the most changes were almost twice as likely to get ill as the 10 percent who had experienced the least. However, I am able to find only a few books that speculate on the relationship between economic instability, stress, creativity, and illness: George Pickering's *Creative Malady* (1974), Totman's (1979) *Social Causes of Illness*; and M. Harvey Brenner's *Mental Illness and the Economy* (1973).

If indeed there is further research that confirms the relationship between stress and illness, the scope of my model can be expanded to suggest that when circumstances suddenly change, we really do not know in what direction the body may "gamble," or which part of the body might be affected by the change.

In conclusion, if my views are correct, the ancient proverb "Sane mind, sane body" can be paraphrased thus: "Secure mind, secure body." I do not know what the word "sanity" means—it seems to be defined by custom alone.

4. What Have Other Writers Said about Creativity and Customs?

Many writers from ancient to modern times have implicitly associated "thinking" and "creativity" with suffering, loss of status, and some disorder. Moreover, some of them have advocated order, recognizing that the price paid for it is diminished creativity. Some of these writers were mentioned in my first book; this brief section complements that picture by quoting others.

Aristotle in the *Politics* objects to the proposal of Hippodamus for awards for new ideas. In a *settled* regime he considers them suspect. For "a citizen will receive less benefit from a change in the law than damage from becoming accustomed to disobey authority . . . the law itself has no power to secure obedience save the power of custom, and that takes a long time to become effective" (pp. 82–83). Oscar Wilde echoed this view when he said in "The Soul of Man Under Socialism" that "It is through disobedience that progress has been made."

Adam Smith in a lesser-known essay, "The History of Astronomy," wrote explicitly that men "have seldom had the curiosity to inquire by what process of intermediate events [a] change is brought about. Because the passage of the thought from the one object to the other is by custom become quite smooth and easy" (pp. 44–45), and that "it is well known that custom deadens the vivacity of both pain and pleasure, abates the grief we should feel for the one, and weakens the joy we should derive from the other. The pain is supported without agony, and the pleasure enjoyed without rapture; because custom and the frequent repetition of any object comes at last to form and bend the mind or organ to that habitual mood and disposition which fits them to receive its impression, without undergoing any violent change" (p. 37).

Similar statements have been made by poets and writers and not just philosophers. Wordsworth wrote that "Wisdom is offtimes nearer when we stoop; than when we soar"; Jean-Paul Sartre stated that "Genius is not a gift, but rather the way out one invents in desperate situations"; and Proust, in *Remembrance of Things Past*, wrote that "Everything great in the world comes from neurotics. They have composed our masterpieces. We enjoy lovely music, beautiful paintings, a thousand delicacies, but we have no idea of their cost, to those who invented them, in sleepless nights, spasmodic laughter, rashes, asthmas, epilepsies . . .", and in the chapter "The Past Recaptured" he notes that "Happiness is beneficial for the body, but it is grief that develops the powers of the mind." One can also find this opinion expressed in Samuel Johnson's statement that "the mind is seldom guided to very rigorous operations but by pain or the dread of pain," and, sarcastically, in Anatole France's *The Revolt of the Angels* (the epigraph of one of the sections of this chapter). John Dewey argued long ago in *Human Nature and Conduct* that thinking was an adaptive behavior triggered by unfulfilled expectations, while William Faulkner said more than once that he created characters in violent circumstances in an effort to get at the truth of the human heart. One can also find this viewpoint, expressed differently, in the writings of Isaiah Berlin (1981). In the essays on Moses Hess, Marx, and Disraeli, he suggests that only when one belongs to a community can one manage a full life undistorted by neurotic self-questioning about one's true identity and be free from feelings of inferiority, real or imaginary. Those who do not belong, however, "hit upon various more or less conscious solutions to their problems of self identity." These lead, eventually, to original insights—"a neurotic distortion of the facts," as Berlin puts it. In fact Berlin thinks that many of Marx's and Disraeli's ideas evolved in the first place not as tools of analysis, but as comforting myths to rally oppressed spirits, perhaps those of the authors themselves.

In the more technical language of the management literature some similar views have been expressed by Herbert Simon (1959) (although he admits in his 1957 work that his intention was to build a new vocabulary rather than a theory):

(a) Where performance falls short of the level of aspiration, search behavior (particularly search for new alternatives of action) is induced.

(b) At the same time, the level of aspiration begins to adjust itself downward until goals reach levels that are practically attainable.

(c) It the two mechanisms just listed operate too slowly to adapt aspirations to performance, emotional behavior—apathy or aggression, for example—will replace rational adaptive behavior. [p. 87].

Simon's first proposition exactly reflects one facet of my model. His second proposition, while plausible, will not always hold true. Some people (call them "entrepreneurs") may turn out to be determinate enough and may gamble on new ideas until they achieve their expectations. But Simon is right, in part, in his third proposition when he states that when aspirations are not achieved one may gamble on criminal acts and that the lack of achievement takes its emotional toll. However, he is misleading in calling the alternative "rational" adaptive behavior. If somebody becomes suddenly poorer, why is it "rational" to adapt to the new situation by just staying poorer, rather than starting to gamble on new ideas and trying to restore one's position in the distribution of wealth? It should be emphasized that if one defines the "search behavior" (described in Simon's first proposition) as "rational," one must also define the aggressive one as such, since both types stem from the *same* trait, a trait precisely defined in the model. But there seems to be little doubt that Simon's view of human nature bears some strong resemblance to mine (although his vocabularly is different). In his book with James G. March he writes:

> . . . we may conclude that high satisfaction, per se, is not a particularly good predictor of high production nor does it facilitate production in a causal sense. Motivation to produce stems from a present or anticipated state of discontent and a perception of a direct connection between individual production and a new state of satisfaction. [p. 51]

The definition of the entrepreneurial act within the model gives precision to these words.

Finally, Shils's (1981) observation on the linguistics associated with this subject should be quoted. He notes that 'Originality' in its first usage in the English language did not describe the working of man's innate genius. It referred, rather, to the greatest of all burdens of the past, 'original sin'. . . : the 'fall' of man from his state of grace and his consequent expulsion from the Garden of Eden" (p. 152). "Originality," which now has, in some societies, a rather positive connotation, thus initially (and unsurprisingly) had a rather ambiguous one: after all, tasting from the "tree of knowledge" and opening the box (Pandora's) led to criminal acts, too. Shils also notices that a "genius" originally meant a demon which possessed a human being and made him perform acts beyond the power of ordinary human beings, while Plato noted that during the creative act people are out of their minds—how this statement receives a literal interpretation with mathematical symbols is the surprise of the next Appendix.

Appendix 1.2:
On Making Up Our Minds . . .

Why Do People Take Risks?[34]

The model presented below is based on Brenner (1983b); its major features are summarized here. Let us assume that people's behavior is relative, that is, their utility function (or their "satisfaction") $U(\,\cdot\,,\,\cdot\,)$ depends both on their wealth, W_0, and on the percentage of people whose wealth is greater than W_0, $\alpha\ (W > W_0)$. With reference to international decisions, the comparision will refer to the wealth of neighboring nations ("neighboring" as defined by the available military technology); more about these comparisons later. Let us use $\alpha_1(W > W_0)$ to denote the percentage of people who are richer within the nation, and $\alpha_2(W > W_0)$ to denote the percentage of people who are richer, but who belong to other, neighboring nations.

Assume that an individual's position in the distribution of wealth, national or international, is that which was expected. His utility function, then, can be formally written as:

(1) $\qquad U = U(W_0, \alpha_1(W > W_0), \alpha_2(W > W_0) \,|\, \alpha_1(\,\cdot\,), \alpha_2(\,\cdot\,))$

which means that his behavior is relative to his position in both the domestic and the international distribution of wealth, given that his realized positions are the ones expected. It is assumed that, holding other people's wealth constant, one's utility increases when one's wealth increases:

(2) $\qquad U_1 = \dfrac{\partial U}{\partial W_0} > 0$

where U_1 denotes the marginal utility of wealth. It is also assumed that an increase in *the percentage* of people (either local or foreign) whose wealth is greater than one's own decreases one's utility:

(3) $\qquad U_2 = \dfrac{\partial U}{\partial \alpha_1} < 0$

$\qquad\qquad U_3 = \dfrac{\partial U}{\partial \alpha_2} < 0$

U_2 and U_3 represent the change in one's utility when one's position in the distribution of wealth changes.

The sign of U_1 represents the usual assumption that the marginal utility of wealth is positive. The second and third assumptions imply that when one's relative wealth diminishes, although one's absolute wealth stays constant,

one's utility diminishes. On the level of everyday behavior one may understand this characteristic as being due to one's envy, ambition, or fear of
being hindered by internal opposition. When the comparison with foreigners is made, this characteristic may reflect the fear of being hindered by
a foreign power. As shown below, these characteristics of human behavior
lead people to gamble on a wide range of activities: on innovations, on going
to war, on revolutionary ideas. At the same time, this characteristic also
provides the incentives for the relatively richer to insure themselves. The
fact that the model leads to the aforementioned implications led me to
interpret the "utility" function as representing a probability of a kind of
survival (see Brenner 1983b). In order to familiarize the reader with the
mechanism of this model, the analysis starts with a very simple problem of
personal initiative: the decision to gamble on a lottery and to take out
insurance. For simplicity's sake, it is first assumed that such decisions are
taken without reference to the wealth of foreigners—i.e., $\alpha_2(W > W_0)$ is
assumed to stay constant, and is thus dropped as an explicit variable from
the utility function. After presenting these implications of this model, those
concerning "foreign" affairs are noted.

Planning Gambling and Taking Out Insurance

Let us show that starting the analysis from this utility function, and the
assumption that the individual seeks to maximize his expected utility, it is
possible to derive testable implications as to who will engage in a gamble
without making strong assumptions as to the shape of the utility function:
the utility function may be linear in its two components.

Suppose that the individual possesses wealth, W_0. This represents the
amount one expects to own by engaging in customary behavior. The percentage of the population whose wealth is greater than one's own is now
simply denoted by $\alpha(W > W_0)$. Assume that the individual is faced with a
gamble in which he has a probability, p, of winning a large amount of
money, H, and a probability, $1 - p$, of losing a smaller amount of money, h,
the price of the lottery ticket. What is meant by "large" amount of money is
a sum sufficient to change the individual's place in the distribution of wealth.

If the consumer wins the gamble, his utility will be $U(W_0 + H, \alpha(W > W_0 + H))$. If he loses, his utility will be $U(W_0 - h, \alpha(W > W_0 - h))$. The
individual engages in the gamble if his current utility is less than the expected
utility he would have if he gambled. Formally, this statement means that:[35]

(4) $\quad p\, U(W_0 + H, \alpha(W > W_0 + H))$

$\quad\quad + (1-p)\, U(W_0 - h, \alpha(W > W_0 - h)) > U(W_0, \alpha(W > W_0))$

In order to simplify the notation, I have omitted the conditional statements.
Assume that $U(\cdot, \cdot)$ is linear in both of its components, that is:

(5) $\quad U = a\, W_0 + b\, \alpha(W > W_0) \quad\quad a > 0, b < 0$

Then from (4) one obtains:

$$(6) \quad pa\,W_0 + paH + pb\alpha(W > W_0 + H) + (1-p)a\,W_0 - (1-p)\,ah$$
$$+ (1-p)b\alpha(W > W_0 - h) > aW_0 + b\alpha\,(W > W_0)$$

Assume that the gamble is fair, i.e., $pH = (1-p)h$. Then from (6) obtains that in order for an individual to participate in a gamble,

$$(7) \quad pb\alpha(W > W_0 + H) + (1-p)\,b\alpha(W > W_0 - h) > b\,\alpha(W > W_0)$$

Since $b < 0$, one obtains:

$$(8) \quad p\,\alpha(W > W_0 + H) + (1-p)\,\alpha(W > W_0 - h) < \alpha(W > W_0)$$

If h, the price of the lottery ticket, is small relative to W_0 and H, so that losing it does not change one's position in the distribution of wealth (a \$5 ticket could hardly do that when there are 200 million people in the economy), then $\alpha(W > W_0 - h) \simeq \alpha(W > W_0)$,[36] and from (8) one obtains (the approximation is made for simplicity's sake only; when the dynamics of the model are later discussed with regard to crime, creativity, and the gamble on new ideas, no approximations are made):

$$(9) \quad \alpha(W > W_0 + H) < \alpha(W > W_0)$$

which implies only that the prize, H, must be so significant that the percentage of people who possess more wealth than $W_0 + H$ is less than those who possess more wealth than W_0 (for pyramidal wealth distributions this condition is fulfilled—see the numerical example at the end of this Appendix). The correct interpretation given to conditions (4) and (9) is that more of the hopes of the relatively poorer are derived from the opportunity to gamble on lotteries.

Would people gamble if there was an equal distribution of wealth in a society, and customs or tax laws existed that *were expected* to maintain wealth at equal levels in spite of fluctuations in the wealth of some because of different abilities or chance? If such customs did exist in a society, then $\alpha(W > W_0) = 0$ and $U = aW_0$ only. Thus, as long as these customs are expected to be enforced, so that any variations in some people's wealth are expected to be either redistributed or compensated for, there is no rationale for the market of gambling to exist. This market emerges, in my model, only when the distribution of wealth becomes unequal.

This conclusion does not imply that in more egalitarian societies people will not play games: they will. Only these games will have entertainment value, rather than value for changing one's position in the distribution of wealth.[37]

Let us show that individuals who gamble may also insure themselves, and for the same reason: an attempt to *prevent* a change in their position in the distribution of wealth. The type of insurance my model can deal with involves amounts that can significantly change one's (or one's offspring's) position in the distribution of wealth, such as life insurance, home insurance, fire insurance, and so forth.

Assume that there is a small probability, p, of losing a large amount, H (because of fire, for example), and a great probability, $1 - p$, of retaining the initial sum (if fire does not occur). The consideration for taking out a fair or an unfair insurance policy at a price, h, which compensates the individual if the unfortunate event occurs, is similar to the condition for the participation in a fair or an unfair gamble (the "unfairness" now representing the insurance premium):

$$(10) \quad U(W_0 - h, \alpha(W > W_0 - h)) > pU(W_0 - H, \alpha(W > W_0 - H))$$
$$+ (1-p) \, U(W_0, \alpha(W > W_0))$$

from which it follows that the same individuals who gamble also insure themselves, and they perform both acts for the same reasons: either to change or to avoid changing their relative position in the distribution of wealth (see the numerical example in this Appendix). Again, the correct interpretation that can be given to condition (10) is that more of the security of the relatively rich is derived from the opportunity to insure themselves. As to the question, when is it more likely that insurance markets will rise?, the answer is that an unequal distribution of wealth, a pyramidal wealth distribution in particular, leads to the emergence of these markets. However, if customs exist whereby people provide one another with a wide range of assurances, insurance will take the form of close ties among members of a group, rather than being taken out in "formal markets."[38]

Without providing a formal model both Galbraith and Schumpeter seemed to base some of their arguments on a view of human nature that bears similarities to mine. Galbraith (1958) wrote that "the notion so sanctified by conventional wisdom, that the modern concern for security is the reaction to the peculiar hazards of modern life, could be scarcely more in error" (p. 109). Galbraith's view is that "with increased well being all people become aware sooner or later that they have something to protect" (p. 107). Schumpeter (1919) wrote:

> It is noncontroversial that the class situation in which each individual finds himself represents a limitation on his scope, tends to keep him within the class. It acts as an obstacle to any rise into a higher class, and as a pair of water wings with respect to the classes below. [p. 163]

This quotation exactly reflects the static part of the model (expressed in more poetic language, "water wings" is a metaphor for insurance). But Schumpeter never discusses how, in what circumstances, and why people will break away: How can one explain both the immobility and the mobility *within* the same model? The next section provides my answer.

The Consequences of Sudden Disorder

Let us turn now to the dynamic part of the model, which is its most significant part.

When one loses one's position in the distribution of wealth one's incentives to gamble increase. The proof is simple: assume that when one's expected wealth was W_0 one did not gamble:

$$(11) \quad U(W_0, \alpha(W > W_0)) > pU(W_0 - h, \alpha(W > W_0 - h))$$
$$+ (1-p) \, U(W_0 + H, \alpha(W > W_0 + H))$$

What is the necessary condition for an individual to gamble on this lottery when his wealth diminishes? Let us assume that the wealth diminishes unexpectedly and significantly, that is, $W_1 < W_0$. Then, in order to take part in the game in which he was previously reluctant to play, the inequality in (11) must be reversed:

$$(12) \quad U(W_1, \alpha(W > W_1)|W_1 < W_0) < pU(W_1 - h, \alpha(W > W_1 - h)|$$
$$W_1 < W_0) + (1-p) \, U(W_1 + H, \alpha(W > W_1 + H)|W_1 < W_0)$$

(For the sake of simplicity and $\alpha(\cdot)$s are omitted from the conditional statements.) Under what conditions concerning the distribution of wealth in the economy will this "taste" for gambling emerge? For both (11) and (12) to hold true the following inequality must also hold true:

$$(13) \quad U(W_0, \alpha(W > W_0)) - U(W_1, \alpha(W > W_1)) > p[U(W_0 - h,$$
$$\alpha(W > W_0 - h)) - U(W_1 - h, \alpha(W > W_1 - h))] + (1-p)$$
$$[U(W_0 + H, \alpha(W > W_0 + H)) - U(W_1 + H, \alpha(W > W_1 + H))]$$

Let $\Delta W = W_0 - W_1$ and let us continue to assume that the utility function is linear in both of its components. We then obtain:

$$(14) \quad a \, \Delta W + b(\alpha(W > W_0 - h) - \alpha(W > W_1))$$
$$> p[a \, \Delta W + b(\alpha(W > W_0 - h) - \alpha(W > W_1 - h))]$$
$$+ (1-p) \, [a \, \Delta W + b \, (\alpha(W > W_0 + H) - \alpha(W > W_1 + H))]$$

Assuming that the gamble is fair (that is, $ph = (1-p)H$), and that h is relatively small and the population large enough, $\alpha(W > W_0) \simeq \alpha(W > W_0 - h)$ or that the society is perceived to have a distinct pyramidal class structure we obtain:

$$(15) \quad (1-p) \, [\alpha(W > W_0) - \alpha(W > W_1)] < (1-p) \, [\alpha(W > W_0 + H)$$
$$- \alpha(W > W_1 + H)]$$

or

$$(16) \quad \alpha(W > W_1) - \alpha(W > W_0) > \alpha(W > W_1 + H) - \alpha(W > W_0 + H)$$

a condition which requires only that as wealth increases the fraction of people in each additional range of H diminishes. In everyday terms this result means that there is a small "upper" class, a larger "upper-middle" class a still larger middle class and lower-middle class, and some fraction of poor, a wealth-distribution pattern that now seems to characterize most countries with large populations.

It should be reemphasized that when one loses part of one's wealth one is more likely to start gambling.[39] This result demonstrates how in this model individuals can "change their minds," and shows as well that the change does not require any statement about changing the "utility function"—the function stays the same;[40] only one's position in the distribution of wealth is unexpectedly significantly altered.[41]

Gambling on Criminal Acts

To commit a crime is risky: one might be caught and punished, or one might get away with it. The sorts of crimes analyzed here will be those against property, where monetary rewards are possible, *or* where destroying the wealth of others, with a significant effect on the distribution of wealth is possible.

Before presenting the predictions that can be derived from this model as to the likelihood of people committing crimes, it is useful to say a few words on what the term "crime" means.

The word does not have a precise meaning: Robin Hood was viewed by some as a criminal, by others as a social reformer. "Terrorist" movements frequently refer to themselves as "liberation armies." Are they made up of "criminals" or "freedom fighters"? It seems that whether something is a considered a crime depends on whether or not people agree with the present distribution of wealth. When the majority accepts the channels by which wealth is redistributed (markets, governments, family ties, and so forth), there is agreement as to what the term "crime" means.

Let us assume that there is agreement in the society as to what constitutes a crime. Who are the people most likely to commit crimes in such a society? Let G be the value of punishment if one commits a crime, and let H be the "prize" if one gets away without punishment. W_0 denotes one's initial wealth and p the probability of being caught. The expected satisfaction from this crime, denoted by EU, is:

$$(17) \quad EU = pU(W_0 - G, \alpha(W > W_0 - G))$$
$$+ (1-p) U(W_0 + H, \alpha(W > W_0 + H))$$

The sum on the right-hand side shows the expected outcome: the first term, of being caught, and the second, of being undetected. Assume that one did *not* commit this crime when one's expected wealth was W_0, that is:

$$(18) \quad U(W_0, \alpha(W > W_0)) > EU.$$

Now let us assume either that one's wealth diminishes to W_1 or one's wealth stays constant, but that everybody else who enters into the definition of $\alpha(\cdot)$ becomes suddenly richer. Then the individual who became relatively poorer is more likely to commit the crime that he previously was reluctant to commit, and the inequality in (18) may be reversed (holding probabilities, punishments, and gains from stealing constant). For the condition for this change of mind is (still assuming the same linear utility function as before, and making the same calculations in (14)):

(19) $\alpha(W_0 > W > W_1) > (1-p)\, \alpha(W_0 + H > W > W_1 + H)$

$$+ p\, \alpha(W_0 - G > W > W_1 - G)$$

The meaning of this condition is clear: $\alpha(W_0 > W > W_1)$ denotes the fraction of people who have suddenly outdone our individual. The term on the right-hand side denotes the fraction of people that, on average, would outdo him if he undertook the risk. If this average is smaller than $\alpha(W_0 > W > W_1)$, then the individual will undertake the risk that he was previously reluctant to undertake. Thus, in this model nothing can be said about the propensity of the "poor" to commit crimes. Those who were suddenly outdone by their fellows are the ones whose propensity to commit crimes increases. Why don't the poor commit crimes? The answer suggested by the model is similar to the one given a long time ago by Knight (as recalled by Gary Becker): *unthinking* obedience, literally speaking.

It should be pointed out that in this case of a gamble on a criminal act, the meaning of the probability, p, is different from the case of lotteries and insurance. While in the case of the latter two a comparison could be made among the probabilities of different people, in this case no such comparison can be made, since the calculations are subjective. That is the reason why the comparison refers to the behavior of the same individual in suddenly altered circumstances. In terms of this model one can thus argue that one's utility function depends on one's "history." The mathematical translation one can give to this word—one's "history"—in this model is that the utility function will depend on the *unexpected* fluctuations that occurred in one's relative position in the distribution of wealth in the past; let us denote them by $\Delta\alpha_t$ (the subscript t indicating the time it occurred). Then the precise way to rewrite (17) is:

(17a) $pU(W_t - G, \alpha(W > W_t - G) \,|\, \Delta\alpha_{t-1}, \Delta\alpha_{t-2} \ldots) +$

$$+\ (1-p)U(W_t + H, \alpha(W > W_t + H) \,|\, \Delta\alpha_{t-1}, \Delta\alpha_{t-1}, \Delta\alpha_{t-2} \ldots)$$

The utility function stays the same if $\Delta\alpha_t = 0$ (i.e., no unexpected changes occur), but it may change following an unexpected change in one's position in the distribution of wealth, when the individual has already learned from his experience. The meaning of the calculations leading to the result stated in (19) is that when a significant, unexpected change occurred in period t the individual may have chosen a new idea, a new course of action during that period, which in period $t + 1$ leads to the revised parameters of the utility function. In this sense one is "learning by doing" in this model. In the next sections, too, the probability p and the utility function should be thus interpreted, although for the sake of simplicity the $\Delta\alpha_i$'s are omitted from the notations.

Perhaps it is also useful to point out in what sense the conditions stated in (18) and (19) represent solutions to a dynamic maximization problem. Formula (18) shows that with W_0 the individual maximized his utility and avoided taking some risky courses of action. Formula (19) shows a characteristic of the new courses of action that the individual may pursue in order to *increase* wealth when his relative wealth has suddenly diminished. This is the way that the goal of the "maximization of wealth" can be interpreted within this model. In contrast, in standard economic analysis, "maximization of wealth" refers to the solution of an *allocational* problem: i.e., how to allocate a given, well-known amount of wealth. But the standard analysis cannot say much as to how that amount of wealth has been obtained to start with.

Creativity and the Gamble on Political Ideas

The wish to huddle in groups, to create noise, and to escape from thought is a sign of desperation and despair.

Montale

Since gambling and crime either redistribute or destroy wealth, the human trait postulated here seems to have no survival value. In order to show that it *does* have such value, it must be shown that the trait postulated here is also *beneficial*. In this section it is shown that this trait leads human beings to offer novel ideas in business, science, arts, technology, and the organization of social institutions (and as shown in Brenner (1983b) it leads them to interpret causal relationships). The supply of novel ideas increases when a group's relative wealth diminishes or the society's total wealth diminish (simultaneously with a change in the distribution). This trait provides the individual or the human race with their means for survival and enables them to maintain "wealth" per capita (on average) at a stable level, although the perception of what constitutes "wealth" changes significantly, and errors are made when interpreting causal relationships.

"Gambling" is an activity that is related not only to games but to *all* activities in which returns depend on luck, rather than on specific skills or already available information. Since this is a characteristic of gambling, betting on an idea for which no empirical evidence yet exists also represents a gamble—although there are differences between this "gamble" and the one embodied in a lottery.

Let P denote the amount invested in developing an idea, which in addition to the direct costs of investment in time and other resources necessary to develop it, also includes the resources invested in trying to estimate the potential demand for the idea and its applications. Let $\alpha(W > W_0)$ again represent the percentage of the relevant population above one's wealth, W_0, and let H denote the increase in wealth that one expects to gain by selling an idea. If unsuccessful, with a probability, p, the potential innovator knows that he will lose the amount, P (which includes losses due to diminished reputation). Translated to mathematical notation, the conditions for an individual to become an entrepreneur, that is, to "bet" on a new idea for which *no* empirical evidence exists with which to test it when he *begins* working on it, is similar to the conditions that lead one to gamble on a criminal act:

$$(20) \quad U(W_0, \alpha(W > W_0)) < pU(W_0 - P, \alpha(W > W_0 - P))$$
$$+ (1-p) \, U(W_0 + H, \alpha(W > W_0 + H))$$

Let me point out the differences between this condition and the one that defines participation in a lottery, in spite of the similarity in the mathematical formulation. Here the value of the "prize," H, differs among individuals: people's evaluation of the potential demand for home computers and the solution to open problems in science are subjective. Also, probability here represents a subjective judgment by an individual, and there is no way to prove that he is right or wrong. Thus one could not say whether or not one is "right" or "wrong" when assigning a value to these probabilities. With lotteries, on the other hand, probabilities are defined in terms of processes that can be repeated many times, and their value will be the same for everybody who plays them.

In spite of these differences between the two types of "gambling," the following conclusion can be drawn: when one's relative wealth drops, one has a greater incentive to "gamble" on novel ideas. This conclusion is thus similar to the one obtained in the previous section, and the proof is exactly the same. Assume that one did *not* gamble on a novel idea, but just followed custom and habit, so that the sign of the inequality in (20) is reversed. When one's wealth diminishes relatively, holding probabilities, costs, benefits, and the distribution of wealth constant, one is more likely to contemplate an idea that one previously dismissed (i.e., one that was latent in one's mind but did not affect one's actions—there was more about this point in the previous Appendix).

Again the similarity with some of Schumpeter's (1927) views should be noted. He wrote that one way to jump classes is "to do something altogether different from what is . . . ordained to the individual," in particular "to become an entrepreneur—which does not, of itself, constitute class position, but *leads* to class position" (pp. 173–74, italics in original). He also notes that

Family and social history show that, in addition to the elements of chance . . . the method of rising into a higher class . . . [is] the method of striking out along unconventional paths. This has always been the case, but never so much as in the world of capitalism. True, many industrial families, especially in the middle brackets, have risen from small beginnings to considerable or even great wealth by dint of hard work and unremitting attention to detail over several generations; but most of them have come up from the working and craftsman class . . . because one of their members has *done something novel*, typically the founding of a new enterprise, something that meant getting out of the conventional rut. [p. 174; italics in original]

Schumpeter frequently refers to the "social pyramid" and suggests that one always needs a symbol to recognize a social class and to distinguish it from other social classes (according to him marriage can serve as such a symbol [p. 141]).

Gambling on revolutionary ideas that advocate a redistribution of wealth can be discussed separately, since the mathematical conditions governing this situation differ from the condition that defines gambling on other new ideas. In this case one individual's behavior affects that of the others: not only do people gamble on one individual's ideas (ideas that represent a fixed point of reference in the midst of growing chaos), but also each individual in the group whose position in the distribution of wealth has been worsened realizes that others in the group are likelier to change their minds. That is the reason why the probabilities in this case will depend on the size of the group whose position in the distribution of wealth has been worsened (in contrast to the cases discussed until now in which one individual's entrepreneurial act was assumed to have no direct effect on the behavior of others).

Let W_0 denote one's initial wealth, when one was reluctant to contemplate gambling on a revolutionary idea:

$$(21) \quad aW_0 + b\alpha(W > W_0) > pa(W_0 - P) + pb\alpha(W > W_0 - P)$$
$$+ (1-p) a W_R + (1-p) b\alpha(W > W_R)$$

where P denotes the punishment and costs of investing in the revolutionary act, and W_R denotes the expected wealth if the revolution succeeds; p denotes the subjective probability given to the revolution's lack of success in catching fire (and recall that a and b depend on past fluctuations).

Assume that one's wealth diminishes but that the distribution of wealth in the society did not change. Then the condition one obtains for the gamble is similar to that of gambling on any new idea. However, if a *significant* percentage of the population who may belong to an organized group loses its wealth, the probability that many people will be ready to change their minds increases. For the change of mind to take place, the inequality in (21) must be reversed. Let $1 - p_1$ denote the subjective probability given to the revolution's success, where $p_1 < p$, and we obtain:

$$(22) \quad a W_1 + b\alpha(W > W_1) < p_1 a(W_1 - P) + p_1 b\alpha(W > W_1 - P)$$
$$+ (1-p_1)a W_R + (1-p_1)b\alpha(W > W_R)$$

For both (21) and (22) to be fulfilled we obtain:

(23) $a(W_0 - W_1) - b\alpha(W_0 > W > W_1) >$

$$> a(pW_0 - p_1W_1) + a(p_1 - p)P +$$
$$+ b[p\alpha(W > W_0 - P) - p_1\alpha(W > W_1 - P)]$$
$$+ (p_1 - p)[a W_R + b\alpha(W > W_R)]$$

where $p_1 - p$ is negative. Since the last term in brackets denotes the expected satisfaction if the revolution succeeds, and $p_1 - p$ is negative, the whole last term is negative. Thus a revolutionary idea which advocates a redistribution of wealth in *favor* of the déclassé group (i.e., diminished $\alpha(W > W_R)$) always has a greater appeal when a group's position is disproportionately worsened, for the probability that the inequality in (23) is reversed increases.

The implications of the model concerning international affairs are examined in the text. But it is useful to note several things. The mechanism described until now refers to a situation in which the distribution of wealth has suddenly been significantly altered. But one may ask: why has it been altered? One answer to this question has been discussed in detail in my previous book, and it will be discussed again below in chapters 3, 4, and 7.

On Dynamic Equilibrium

One may raise the following question: Even if it is true that losing $5 on a lottery ticket will not change one's behavior (and one is thus in "eqilibrium"), what happens if one buys 100, 1,000, or more tickets? When does one stop buying them? The answer to this question is that once the individual has lost a relatively large amount by playing in unfair games of chance (suppose that he started playing because he unexpectedly lost part of his wealth), he will perceive that he cannot restore his position in the distribution of wealth by gambling. He may thus gamble on an entrepreneurial act: either by making greater efforts at work or by committing a crime. These possibilities represent the "dynamic equilibrium" in the model (from the individual's viewpoint).

Finally some more philosophical notes. There are similarities between my point of departure and those of some ancient observers when examining human behavior, although they are expressed in very different language. According to Plato human behavior flows from three main sources: desire, emotion, and knowledge; desire and instinct are one; emotion, courage, ambition another; knowledge and reason another. Within my model human behavior flows from three main sources, too: the utility function (call it "desire"?), emotions (call them "envy," "ambition," which stem from making comparisons with others), and customs (which indeed *are* one's "reason" and "knowledge"). The view of "lucky hits" or "Divine inspiration" may also remind some of Aristotle's view of God, who never does anything: he has no desires, no will, no purpose (at least not that one can understand), and he never "acts." Notice that when the criminal or the entrepreneurial gamble is defined in the previous Appendix, p remains *constant* (p changes for those who decide to imitate, become followers, and thus gamble on *other* people's ideas, when revolutionary ones are discussed). A new act is pursued because one's position in the distribution of wealth has been altered. Aristotle also emphasized the role of customs: he is the first to call it "second nature" (see Durant, 1926, pp. 69–72). The model presented here suggests what "the first" is. The drama between the first and the second nature appears in ancient Greece. The word "anomie" comes from the Greek, meaning without laws or customs. Anyone who lost his "nomos" was to be both pitied and feared (in the fifth century B.C. the basis was set for a positive philosophy of "physis," i.e., the "nature," the qualities of human behavior which are inborn, which do not depend on customs [see Nisbet 1982]). Today the sociologists' theory of anomie is based on the works of Durkheim, who perceived anomie as a consequence of change and of destroyed customs, and who emphasized its dark sides (suicide, among others), but who also noted correctly that the advancement of the arts and crafts of civilization is linked with the increase of unhappiness and the loss of the secure contexts of belief and of membership in well-defined communities. Recently Nisbet (1982) repeated these themes and wrote: "Anomie . . . can be, under certain circumstances, tonic rather than toxic. To a degree, all creative work has implicit a measure of anomie, for the essence of creativity is that the creator move far enough away from one orthodoxy or conventionality, perhaps in conscious approach to another, to have the feeling 'I am I' more strongly than he would were he still closely bound to 'nomos' " (p. 15). The model presented here lends precision to this view of those of the ancient Greeks on customs, nature, and creativity (or, at least, translates them to a more up-to-date language).[42]

Numerical Example

Let us illustrate the calculations in the model by a simple numerical example. Consider a pyramidal perception of the distribution of wealth in which 50% of the population (assumed to be large enough) is expected to own $100,000, 30%, $200,000 10%, $300,000, and the top 10%, $400,000. The utility function is assumed to be the same for everybody:

(1) $U = aW_0 + b\alpha(W > W_0)$ $a > 0, b < 0$

(and recall that a and b depend on past fluctuations in $\alpha(\cdot)$). It is a straightforward exercise to show that the upper class will only insure itself but not gamble, the lowest class will only gamble but not insure itself, while the middle classes will do both. Consider a fair gamble, or a fair insurance, where the possibility exists either to lose or win \$100,000. I must take this sum because it is this sum that can move one from one class to another. Let us make the calculations: consider first the lower-middle class with the \$200,000 wealth. Using the same notation as in the text, the condition for insurance becomes:

(2) $a(200,000 - h) + b\,0.2 > pa\,200,000 + pb\,0.2 +$

$$+ (1-p)a(200,000 - 100,000) + (1-p)b\,0.5$$

which can be reduced to:

(3) $b\,0.2. > pb\,0.2 + b(1-p)0.5 \qquad b < 0$

$$=> 0.2 < 0.5$$

which is thus fulfilled. The condition for gambling is:

(4) $a \cdot 200,000 + b\,0.2 > pa(200,000 - h) + pb\,0.2$

$$+ (1-p)a\,(200,000 + 100,000) + (1-p)b\,0.1$$

which can be reduced to:

(5) $(1-p)b\,0.2 < (1-p)b\,0.1 \qquad b < 0$

$$=> 0.2 > 0.1$$

Similar calculations will show that the richer only insure themselves but will not gamble, since the two conditions are:

(6) $a(400,000 - h) > pa\,400,000 + (1-p)a(400,000 - 100,000)$

$$+ (1-p)b\,0.1$$

(7) $a400,000 < pa(400,000 - h) + (1-p)a(400,000 + 100,000)$

The first is reduced to

(8) $0 > (1-p)b\,0.1$

which is always fulfilled since $b < 0$, while the second never (although with a weak inequality, the richest could be indifferent between participating in a *fair* gamble or not. But they will not participate if the gamble is unfair; the condition for unfair gambles is presented in Brenner 1983b).

This example can also illustrate the dynamic part of the model. Suppose that an individual expected to own \$400,000 and thus only insured himself, but did not gamble. Suddenly he perceives that he cannot reach this level of wealth. He may then start to take part in all types of gambles that he previously avoided, entrepreneurial ones in particular. Consider a gamble where one may lose or win \$100,000, but an individual from the richest class does not participate, i.e.:

(9) $a \cdot 400{,}000 > pa\ 500{,}000 + (1-p)\ a\ 300{,}000$

$$+ (1-p)b\ 0.1 = a \cdot 300{,}000 + pa\ 200{,}000 + (1-p)b\ 0.1$$

(10) $=> 100{,}000\ a > pa\ 200{,}000 + (1-p)b\ 0.1$

Now he suddenly loses \$100,000. What are the conditions for this individual to undertake this particular gamble (notice that probabilities here are subjective and *no* comparisons can be made between the reaction of this individual and that of one whose wealth was expected to be \$300,000 to start with):

(11) $a\ 300{,}000 + b\ 0.1 < pa \cdot 400{,}000 + (1-p)a\ 200{,}000$

$$+ (1-p)b\ 0.2$$

(12) $=> a\ 300{,}000 + b\ 0.1 < a \cdot 200{,}000 + pa\ 200{,}000$

$$+ (1-p)b\ 0.2$$

(13) $=> 100{,}000\ a > pa\ 200{,}000 - b\ 0.1 + (1-p)b\ 0.2$

Conditions (10) and (13) imply that:

(14) $pa\ 200{,}000 + (1-p)b\ 0.1 < pa\ 200{,}000 - b\ 0.1 + (1-p)b\ 0.2$

(15) $=> -b\ 0.1 + (1-p)b\ 0.1 > 0$

$$=> b\ 0.1\ (1-p-1) > 0$$

$$b\ 0.1 \cdot (-p) > 0$$

which holds true since $b < 0$. Of course, for nonlinear utility functions or nonpyramidal wealth distributions these results may not hold true. However, one should note that it is meaningless in this model to play with shapes of wealth distributions, since the question is empirical, namely: what is people's perception of the distribution of wealth? If people's perception is that classes of wealth exist and there is a large "working class," a smaller "lower-middle class," a still smaller "upper-middle class," etc., then it is this perception that must be taken into account in the model. As to the assumption of a linear utility function: the question is why should one complicate assumptions? If I found some facts that were inconsistent with the predictions of the model, I would have to return to the model and change some assumptions. For the moment I am not able to find such facts—so why should I play logical games? What would happen if a peculiar function was found which predicted that the richer gamble more, or that those already born rich are more innovative, though the facts contradict these predictions? (see also the opening section of the previous appendix on this methodological problem; its epitaph is useful, too). Maybe "mathematical economists" (quite a contradiction of terms), who are dealing, by definition, with linguistics and consistency may peer down their nose at such an approach. But my view is that they are out of touch with reality and pursue ideas with no consequences whatsoever in understanding human behavior. They may be interested in consistency, but why should social scientists care about consistency if it can shed no light on facts? Remember what Oscar Wilde said: "Consistency is the last refuge of the unimaginative"?

Finally two points: first, notice that it is inconsistent to assume within this model that people may know other people's wealth with precision, since "wealth" in this model also depends on one's intentions and opportunities to insure oneself and to gamble. One individual may only vaguely know what other people's intentions and perceptions of wealth are (see also Schumpeter's views quoted earlier in this Appendix). Second, notice that while the sufficient conditions for people to start to commit noncustomary acts recall a second-order condition, they do not necessarily imply such a condition. For a discrete pyramidal distribution of wealth implies that the first derivatives of the utility function are not defined everywhere.

Appendix 1.3:
. . . on Political Thought, in Particular

Since this book deals with various subjects related to the domain of political science and suggests political solutions to some problems, it may be useful to point out briefly how my arguments relate to some classic works in the field of political thought.

Fluctuations in population played a central role in my previous book and play a central role here, as emphasized in chapters 3, 4, and 7. Sudden, large fluctuations destroy customs (including religious ones) and lead to a redistribution of wealth. Then, through the process outlined in this chapter (and in more detail in Brenner 1983b), new forms of political organizations emerge: these organizations attempt to substitute for the abandoned, forgotten customs and to restore and maintain stability. Since such fluctuations occurred more frequently in Western Europe than in other places, one would expect that "political thought" would thus play a greater role there than elsewhere. Indeed, Nisbet (1973) noticed that:

> Of all communities and continuities of social thought in western European history, the political is the most distinctive in kind and has exerted the greatest effect upon the rest of the world. Other civilizations of the past have of course known political organization, beginning with Egypt and China . . . But it is only in Western civilization, commencing with ancient Greece, that the state came to be conceived, idealized indeed, as a community worthy of a status that elsewhere attended only religion. [p. 1]

When customs break down, the problem faced by a political leader is to decide what strategy he can gamble on to relink the disordered population. A new religious idea is one possibility (notice that the word "religious" originates from the Latin "ligare" which means "to link"). An alternative is that of the secular state. Thus it may be not such a big surprise that Plato, one of the fathers of Western philosophy, developed the second idea in the aftermath of the Peloponnesian War, a period of social breakdown in Athens: his *Republic* argues for a dominant role for political ties (above religious and kinship ones).

It is beyond my purpose to examine Plato's philosophy, since this book deals with practical rather than philosophical issues. However, as in the case of Machiavelli, it may be useful to point out that his view of human nature bears some resemblance to mine. For the benefit of my readers it is useful to quote a philosopher's summary of Plato's views of human nature, since otherwise it may appear that the similarity is forced and that I am presenting a biased view. Here is Nisbet's (1973) summary:

> . . . as background for Plato's philosophy of the political community, there is his notable distinction between nature and convention . . . "Nature" (or *physis* as the Greeks called it) refers to the condition that may be found in either an organism or institution after all the strictly artificial and conventionalized attributes have been stripped from it by one's mind. "Convention" (or "*nomos*") is, by contrast, that which is accidental or superficial, the chance result of time, or place or culture.
> [p. 4; italics in original]

The model summarized in this chapter defines that fundamental "natural" characteristic of human beings. The rest of our characteristics are due to "customs," which emerged in some circumstances by chance: an individual offered an idea on which the rest of the population gambled.

The conclusions one can draw from the model presented here, however, differ from Plato's as to the nature of the government that can maintain order. The conclusions one can reach from both the model and the facts presented here and in the previous book are much closer to Aristotle's which Nisbet also summarizes:

> [Aristotle] did not have an ideal of government so much as he had an ideal of *the relation between government and the social order*. What was important was, not whether government was monarchy, oligarchy, or democracy, but whether the family, private property, legitimate associations, and social classes were able to maintain themselves free of incessant political invasion or domination irrespective of what form of government existed. From Aristotle's viewpoint—and this would be the basic viewpoint of Burke, Tocqueville, and other nineteenth-century pluralists—almost any form of political government was good if it preserved the all-important spheres of autonomy to which each of the major groups and institutions was entitled within the social order.
> [p. 22; italics in original]

The similarity between these views and mine is clear: within my model keeping the distribution of wealth stable, even if unequal, prevents all entrepreneurial activity and any deviation from customary behavior. In principle any form of government may achieve this goal. It is evident from Aristotle's *Politics* that he did not mind inequality at all for he stated that justice is equality, not for everybody, but only for those who are equals, and inequality is just, not for everybody, but for those who are unequal (and he justified slavery on these grounds). For more on Aristotle's views see the two previous Appendixes.

The difference between the views and styles of Aristotle and Plato may be explained, in part, by the fact that Aristotle wrote during more tranquil times than Plato. Plato sounds like an obsessed preacher, pushing his views to extremes, while Aristotle's voice is more distant and relaxed. The different pressures their times put on them may also explain why Plato may have perceived just one strategy for restoring order, while Aristotle, with the insights that only the passing of time can provide, perceived that order can be (and should be) maintained not only by political ties, but by kinship, religious, and cultural ties, too. Durant (1926) shares this view: "what Plato lacks above all . . . is the . . . sense of flux and change: he is too anxious to have the moving picture of this world become a fixed and still tableau. He loves order exclusively . . . " (p. 45). On Plato's views and on the origins of the revolutionary impulse Waelder's (1967) observations should be noted: "The revolutionary impulse may well be said to stem from *frustration*. In a sense, this can be said of all human action which aims at changing the *status quo*. Plato said in his mythical language that *Eros* (Desire) is the child of *Poros* (wealth) and *Penia* (poverty); it moves from the latter toward the former. Desire, in this sense, grows out of frustration . . . " (p. 233; italics in original). He defines "frustration" thus: "[it] is experienced relative to a standard that varies with time, place, and circumstance" (p. 234). These views bear similarities to mine, since one gambles on new acts when one perceives a worsened position in the distribution of wealth (call it a "desire that grew out of frustration"), and these acts can be defined as "new" relative to customs (i.e., "standards that vary with time and place").

Nisbet (1973) also interprets Hobbes's advocacy of centralized power, of ruthless attitudes toward conflicts as a mirror of Hobbes's times. He summarizes Hobbes's views as follows: "We have already seen the capacity of ages of bitter social and ideological conflict for producing works in which centralized power is exalted" (p. 26).

Finally, a brief comparison between Rousseau's well-known views on customs and state and a somewhat less well-known view from the most popular books of all times—the Bible. The freedom of individuals preoccupied Rousseau, the word "freedom" meaning for him freedom from customs, rather than from the legislation of governments (in sharp contrast to some viewpoints today, which, in accord with Aristotle's views, perceive that some balance must exist between the various methods of maintaining stability). He was a harsh critic of the traditional institutions of the church and the extended family, considering them the source of disorder. While the word "free" is used in the next quotation in a way similar to that in which Rousseau used it, the conclusions are exactly the opposite. Here is the biblical description of the behavior of the Jews in Sinai—notice how the word "freely" is used:

> And the mixed multitude that was among them fell a lusting, and the children of Israel also wept again and said: Who shall give us flesh to eat? We remember the fish which we did eat in Egypt freely; the cucumbers, and the melons, and the leeks, and the onions, and the garlic. But now our soul is dried away, there is nothing at all besides this manna before our eyes. [Numbers 11:4–6]

Various interpretations have been given to the word "freely" (*hinam*) in this text. Some took it literally (Iber Ezra, Nahmanides, Abravanel) and interpreted the word either as "relatively cheap," or as indicating a custom among fisherman that allowed fish to be caught in the nets without payment. However, the sages' interpretation was different:

> When they said *hinam* ["freely"] they meant "free of mizvot" [Hebrew word for customary obligations]. Not food or drink, fish or cucumbers, whether given away or cheap, fresh or stale really concerned them, but that freedom from the irksome demands of civilization and standards of self-discipline which they had enjoyed in Egypt . . . But when the Israelites went forth from slavery to freedom, another bondage was imposed on them, more difficult and majestic in its awesomeness—the yoke of Torah and *mizvot* imposed on them in Sinai—self-discipline in the life of the community and individual, in family life and relations with neighbours, on workdays and restdays . . . this yoke of freedom appeared to those accustomed to slavery, as burdensome and irksome. [Leibowitz 1980, pp. 101–2]

The sages argue that accepting customs can make individuals "free," customs thus being viewed in a positive light. Why are customs sometimes perceived in a positive, while at other times in a negative, light?

Notice that the Jews gambled on adopting strict, new customs when they became relatively isolated in Sinai and perceived themselves as being worse off. When leaving Egypt they were composed of a "mixed multitude" that had to be unified by their leader. In contrast there were other times (during the period that Rousseau was writing) when populations were becoming larger, less isolated, in more frequent contact with people from various backgrounds, circumstances that are more likely to lead to a breakdown of customs, including religious ones. Uniform legislation can substitute for this breakdown and restore stability (more on this point can be found in Brenner 1983b, chaps. 2–4). In such circumstances political power, rather than customs, is more likely to be perceived in a positive light.

2 On Gambling, Social Instability, and Creativity Now . . .

Sit down before fact as a little child, be prepared to give up every preconceived notion, follow humbly wherever and to whatever abysses nature leads, or you shall learn nothing.

Thomas Henry Huxley

1. Predictions on Gambling

In spite of a long history of universal condemnation by church and other moral authorities, participation in games of chance has been a widespread phenomenon since ancient times. Ancient Ethiopians, Greeks and Romans, Normans, Gauls, and early Hawaiians gambled.[1] Today members of many societies, regardless of their ideological backgrounds, gamble on a wide variety of events, from horse races to soccer games. In communist countries both lotteries and gambling on soccer games seem to be very popular. In the United States state lotteries are now being reintroduced at a rapid rate, after being outlawed in the 1880s.[2] In Canada, too, the 1970s have seen the introduction of lotteries by provinces as well as by the federal government[3] (at first they were attractively labeled in Quebec as a "voluntary tax").[4]

Are gamblers coming from a particular social class? As shown in the first chapter one participates in games of chance where there is a possibility of winning a big prize either because one is relatively poor, or because one has suddenly become poorer. One can make the following predictions based on that observation as to the motivation for gambling:

— that a chance to win the big prize will be one of the main reasons for lottery-ticket buying. Also a distinction should be made between games of chance that have no entertainment value (such as lottery tickets)[5] and those that do have such value (playing bingo, making new acquaintances, playing cards with friends, playing roulette at Monte Carlo); putting all games of chance in one category may blur one's analysis rather than shed light on the motivations behind gambling.

— That the relatively poor will plan to spend a greater fraction of their wealth on lotteries than the relatively rich, and that people of all classes (upper, middle, lower), who have not previously gambled, may decide to do so when they suddenly lose part of their wealth (for example, when

The coauthor of this chapter was Gabrielle A. Brenner.

they become unexpectedly unemployed). And the contrary: gamblers may suddenly stop gambling or gamble less if they win a big prize. If the facts are found consistent with these predictions, one will be able to conclude that people gamble because they are poor or have become poorer, and not because they have a peculiar, destructive bent.

— That older, rather than younger, people (with the same household income) will gamble. The reason is that a $15,000 income for a fifty-year-old is a different indicator of wealth from the same income for a twenty-year-old. The older man, knowing that he has passed the peak of what he can earn from his traditional occupation, but still wanting to leave an inheritance to his children, is poorer than his younger counterpart with a similar income. But gambling can still make him rich.

— People who have more children will tend to gamble more.[6] The rationale behind this prediction is that for a person with one child a $15,000 income assures a higher position in the distribution of wealth than it does for one with four children: the greater the number of children, the poorer the family (for the same household income).[7]

These predictions show why many empirical researches regarding gambling have been misleading. The researches have gathered information on the income of the buyer or his household without taking age or family structure into account. At times, information on income only was sufficient to reach the conclusion that lower-income groups tend to gamble more.[8] At other times, the incomes of the gambling population seemed too high to support this claim. However, as the above makes clear, income may provide very biased information about one's position in the distribution of wealth, unless some further analysis is made.

The next section presents evidence on participants in games of chance in Canada, Sweden, and England which complements the data presented in Brenner (1983b). All of the evidence seems to support the predictions made above.

Gambling—Some Facts

No games of chance were ever founded on ideological grounds.

Gamblers make a clear distinction between games of chance in which there is a possibility of winning relatively big prizes and games in which the prizes are relatively small. A report of the Royal Commission on Gambling carried out in the United Kingdom in 1951, a 1977 market-research study for Loto Québec, the Report on Gambling in the U.S. (1976), as well as many market-research analyses carried out for the state-operated lotteries in the U.S.,[9] all reveal that the expectation of a large prize is the main reason given by people for purchasing lottery tickets. In contrast, games of chance like bingo, where the physical participation of the gambler during the game is required and the maximum prize is in the range of $5,000, are perceived as having "entertainment value" and are considered "social events."[10]

Table 2.1 presents the incomes of lottery-ticket buyers in Quebec.[11] This table gives the percentage of people in each income category who claim to be regular or occasional buyers of tickets in one of three types of lotteries

Table 2.1 Regular or Occasional Purchases of Lottery Tickets by Income of Purchaser, 1976 Annual Income (% of people polled in each category)

Type of lottery	Annual income, 1976	$0–5,000	$5,000–8,000	$8,000–10,000	$10,000–15,000	$15,000–25,000	over $25,000
Inter-Loto		41	48	52	63	52	59
Super-Loto		28	36	44	41	40	47
Loto-Perfecta		11	26	23	27	23	28

Sour: Sylvestre (1977), vol. 3, tables 3b, 4b, 5b.

(Inter-Loto, Super-Loto, and Loto-Perfecta, each offering a big prize).[12] For instance, 63% of the people whose annual income is between $10,000 and $15,000 are occasional or regular buyers of Inter-Loto tickets. Though these incomes may be average for a young, childless worker they are low for an older worker with four children. As we have no information as to the age or the number of children, however, we cannot conclude whether these are low or average incomes.

In order to show that the buyers are relatively poor, we have looked at both their education and their ages. The reason for looking at education is that education is positively correlated with income. Thus, if we find that the lottery-ticket-buying public is relatively less educated we can conclude that relatively poorer people bought the lottery tickets.[13] The data on income alone does not give us a full enough picture.

Table 2.2 shows that 58% of the people polled with 0 to 7 years of

Table 2.2 Regular or Occasional Purchases of Lottery Tickets by Years of Schooling (% of people polled in each category)

Type of lottery	Years of schooling	0–7	8–12	13–15	over 16
Inter-Loto		58	53	42	39
Super-Loto		43	38	33	24
Loto-Perfecta		25	20	21	11

Source: Sylvestre (1977), vol. 3, tables 3b, 4b, 5b.

schooling, 53% with 8 to 12 years of schooling, 42% with 13 to 15 years of schooling, and 39% with more than 16 years of schooling have answered that they are either regular or occasional buyers of Inter-Loto tickets. Similar trends exist for the other lotteries, the respective percentages being 43%, 38%, 33%, and 29% for Super-Loto and 25%, 20%, 21%, and 11% for Loto-Perfecta. Thus, as predicted, the more schooling people have, the less likely they are to buy Loto tickets.

Table 2.3 shows the percentage of lottery-ticket buyers in each age group

Table 2.3   Regular or Occasional Purchases of Lottery Tickets by Income of Purchaser (% of people polled in each category)

Type of lottery	Age less than 20	20–24	25–34	35–44	45–54	over 50
Inter-Loto	12	33	46	58	68	59
Super-Loto	11	23	32	43	47	46
Loto-Perfecta	12	12	20	24	25	18

Source: Sylvestre (1977), vol. 3, tables 3a, 4a, 5a.

polled: the percentage for the three kind of lotteries is highest for the age groups 45–54 and 55 and older. Considered alone, this finding may have two explanations:

1. These are the ages when people are at the peak of their earning power in the labor market. Thus these data may imply that the richer the individual, the more lottery tickets he will buy.

2. Alternatively, we could say that only when one attains these ages does one realize both that one is not likely to become rich through the labor market, and that the only remaining way to become rich is by buying lottery tickets.

But considering the previous data indicating that the people with less schooling bought relatively more lottery tickets, we can conclude that explanation 1, which is in tune with our theory, seems more valid than 2.

More data on which our views can be tested are given in table 2.4, which shows the percentage of people polled in each age, schooling, and income group who answered that their purchases of lottery tickets are planned. These data, too, show that the people who plan their purchases are people who have had less than seven years of schooling, whose age is above 55, and whose annual income is between $5,000 and $8,000. Thus, as expected, the lottery-playing population in Quebec tends to be older and poorer than the rest of the population.[14]

A similar conclusion can be reached when one examines data on Swedish people who bet on soccer games. While soccer is, theoretically at least, different from a lottery in that it may require both knowledge of the game

Table 2.4   Planned Purchases of Lottery Tickets (% of people in each category who prefer to plan their purchases of lottery tickets)

I. *By age*

20–24	25–34	35–44	45–54	over 55
37	38	35	37	46

II. *By years of schooling*

0–7	8–12	13–15	16
47	38	29	29

III. *By level of annual income* (1976)

$0–5,000	$5,000–8,000	$8,000–10,000	$10,000–15,000	$15,000–25,000	over $25,000
35	60	29	40	43	17

Source: Sylvestre (1977), vol. 3, tables 13a–15b.

and the collection of relevant information about the teams, the difference disappears if one assumes that all the participants in the betting have access to the same information (that is, if no "behind-the-scene" deals among teams are made). In that case deviations from the expected results will be random. Thus participations in soccer pools and lotteries can be compared. It should be also noted that the Swedish example fits the assumptions of the model since there was a possibility of winning a big prize.[15]

The information is derived from answers to questionnaires given to 812 men only between the ages 18–55 (see Tec 1964). Since the tables that have been computed from these answers relate participation in the game to the incomes of the participants, but mention neither ages nor family structures, no useful insights can be obtained from them. However, when the classification has been made according to occupations one observes that 60% of the lower-class respondents are regular bettors compared to 45% within the middle class and 40% within the upper class (see Tec 1964, p. 47). When one relates class background to habitual gambling one finds that among those who come from upper-class homes 38% gamble regularly, from middle-class homes 46%, and from lower-class origins 61%.[16] In addition, one finds that of those whose parents own a business 43% gamble regularly, compared with 54% of those whose parents do not own such property, and that 43% of those who have had higher education gamble regularly compared with 55% of those who have only finished grammar school. As to the age composition: besides the fact that the sample is biased (since people more than 55 years old have been arbitrarily eliminated), the way the tables have been put together enables the reader to learn only that 54% among those between the ages of 18–24 gamble regularly, while 58% of those between 25–34 do so.[17] Thus, as expected, the relatively poor participate disproportionately in games of chance in Sweden, too. As to evidence on age and family structure: in addition to the already-mentioned Canadian data, only evidence from the U.S. could be found, and it is presented in Brenner (1983b).

The data presented until now show that poorer and older people plan to spend a greater fraction of their incomes on lottery tickets than do richer people. But one further prediction of our theory was that if one's realized wealth is less than expected because of illness, the loss of one's job, or any other reason, one is more likely to buy a lottery ticket. This purchase, a reaction to a sudden loss, will not be a regular planned one but rather a spontaneous one. But this kind of sudden illness or accident may happen to anybody, either rich or poor, young or old, with or without schooling. Bad Luck may strike any category of people. Thus, if one puts together the two categories of people—the poorer who plan, and the unfortunates who may decide suddenly to buy lottery tickets—the data may be misleading. Just how misleading depends on the percentage of people who decide to buy *spontaneously* relative to the percentage who *plan* to buy the lottery tickets.

Table 2.5 shows that in Quebec, more than 50% of buyers for each kind

Table 2.5 Planned and Spontaneous Purchase of Lottery Tickets

	Loto Canada	Inter-Loto	Super-Loto	Loto-Perfecta
Planned	275	273	273	272
Spontaneous	380	375	372	373

Source: Sylvestre (1977), vol. 4, tables 1–5, series 9.

of lottery have answered that they buy the tickets spontaneously rather than plan to buy them. This evidence suggests that only less than 50% of the lottery-ticket buyers may be poor. The other 50% may be young, may have no children, but may have just been subjected to an unexpected, unfortunate event (losing a job, not getting the one expected, being suddenly taken ill, and so forth). Anecdotal evidence on this point was revealed in a recent (1982) poll in Quebec (where unemployment had reached an all-time high of 15%). People reported that the money they had previously allocated to beer and wine was now allocated to lottery tickets. The fact that the evidence is consistent with our predictions, in spite of the inappropriate aggregation, suggests that if one could separate the planners from the random buyers, better results could be obtained. Qualitative evidence which does not make an aggregation of the two groups appears in Tec's (1964) study, which concludes that in Sweden gambling behavior is well correlated with a gambler's dissatisfaction at work, and in Brunk's (1981) study on U.S. data, which concludes that dissatisfaction with current income is a strong reason for lottery ticket buying. But what precisely does "dissatisfaction" mean? One can become "dissatisfied" when one's realized position in the distribution of wealth is less than expected. Thus Tec's and Brunk's evidence may be consistent with the predictions of the model.

Further evidence on the relationship between gambling and changes in wealth comes from the Gallup polls (1972) and the work of Cornish (1978), who found that when gamblers win a big prize they spend the money on home-centered items, and from Downes et al. (1976) who found that in horse-race betting small wins are rebet more often than large ones, and rebetting itself is largely confined to regular punters—though of these, three times as many winners save their prizes or spend them on household goods as rebet them. This evidence, as well as the previous data on "dissatisfaction" and gambling, is important, since it suggests that people do not gamble because they are obsessed with this activity. Rather, they gamble in order to become richer. Once they succeed significantly, they spend the money on goods and not on additional gambles. And, on the contrary, they start gambling when their expectations are not realized.

What Have Others Said About Gambling?

Economists, sociologists, and psychologists have examined gambling, and many writers have told gripping stories about gamblers, Dostoyevski being probably the most famous.

Economists have viewed gambling as a matter of "taste" and have labeled it "risk-loving." Many of them have condemned the act (for a summary of their views see Brenner and Brenner 1981). In spite of the fact that some economists (Friedman and Savage 1948) have argued that the "taste" for gambling may be related to the pattern of wealth distribution in the society, they have never been precise about their views. Thus the economic theories of gambling seem to be abstract mental exercises that cannot be contradicted, since gambling is viewed in these theories as purely a matter of taste.

Freud had a rather peculiar view of gambling: he wrote that a gambler was a man who, because of his death wish toward his father, had developed guilt feelings. In order to punish himself he gambles, unconsciously wishing to lose.[18] This seems to be a rather strange theory (and one that can hardly be falsified), but it should be noted that, in a letter to Theodore Reik, Freud ultimately discarded it.[19] Psychoanalysts who devoted attention to gambling either related it to the Oedipus complex or, when trying to bring the subject more down to earth, analyzed only the behavior of compulsive gamblers (see Bergler 1957, Herman 1976). Psychiatrists Darrell W. Bolen and William H. Boyd (1968) summarize the theories of gambling in the psychoanalytical literature in these words:

> Reduced to the most abstract components, gambling is forbidden as an unconscious transgression because of the indirect, intrapsychic satisfaction of multiple aggressive and libidinal determinants. Here is the origin of the abundant pervasive guilt feelings which plague the gambler, as well as the origin of anticipated, desired punishment in the form of gambling loss . . . [p. 626]

This quotation seems to suggest, in very technical words, that in the psychological literature on gambling nothing new has been added to Freud's initial views, and no reference is made to the fact that he discarded them.[20]

In contrast to the psychoanalytic and economic literature in which no attempts have been made to relate gambling to social conditions, sociologists have done just that. Robert Merton's (1957) theory comes closest to the views presented here since he examined "the reference-group theory" in great detail, and two of his followers, Devereux (1950) and Tec (1964), have applied his arguments to gambling. Their viewpoint is that when conventional avenues for social mobility are closed, people will find nonconventional ones, which may include crime and gambling. The main difference between our approach and that of the sociologists is in the precision given to the mechanism that motivates individuals in a society. In fact, one can argue that the approach presented here combines the paradigm of economics (of "utility" maximization) with the sociologists' assumption as to the role of reference groups in shaping human behavior. This combination both provides a falsifiable dynamic theory of gambling, and of risk-taking in general, and gives to the utility function a clear, intuitive meaning.

Is Gambling "Good" or "Bad"?

Ashton's *History of Gambling in England* (1898), Chafetz's *History of Gambling in the United States* (1960), and Sullivan's *By Chance a Winner* (1972) are full of stories of individuals who gambled. At times the stories are sad, at other times exciting. Yet these histories can hardly be viewed as attempts to understand "gambling": the authors have selected *interesting stories*. But "interesting" stories are not provided by the lives of "moderate" or "average" people. Thus Ashton's, Chafetz's, and Sullivan's "histories" can mainly be viewed as records of a tiny minority of *compulsive* gamblers and of the many types of games that people have engaged in throughout history.[21] Perhaps this selective memory provides one of the reasons for the negative attitude of many people toward gambling, since they attribute to all gamblers the characteristics of the few who are compulsive. Yet the facts presented below show that to form opinions based on such extrapolations would be misleading.

The Royal Commission on Gambling in the United Kingdom (1951) concluded that "generally speaking the average expenditure on gambling must be considerably less than the average expenditure on other indulgences such as alcoholic liquor or tobacco" (pp. 49–50) . . . "The great majority of those who take part in gambling do not spend money on it recklessly and without regard to the effect of their expenditure on the standard of living of themselves and their families" (p. 53) . . . "We find no support for the belief that gambling, provided that it is kept within reasonable bounds, does serious harm either to the character of those who take part in it or to their family circle and the community generally" (p. 45).

Similar conclusions were reached by Tec (1964) when she examined the gambling population in Sweden: gambling did not interfere with attempts to advance through conventional channels of social mobility; rather it seemed to provide an additional strategy that served this goal, and when gamblers and nongamblers were compared, neither their intentions to establish business nor their participation rates in training to improve their jobs, differed.[22] Nor did Tec (1964) find any relationship between gambling and crime, marital instability, or the degree of participation in community activities. In fact, she found that gamblers participated in adult-educational courses more than nongamblers did (41% of gamblers versus 33% of nongamblers in the same age groups).[23]

Our research on the U.S. gambling population revealed similar patterns (see Brenner 1983) as did that of the already mentioned Royal Commission on Gambling in the United Kingdom, which concluded: ". . . Whatever the extent of gambling in this country, we have been unable to find any conclusive evidence to support the view that it interferes seriously with production" (p. 40) . . . "The conclusion we have reached on the whole from the evidence, is that gambling is of no significance as a direct cause of serious crime, and of little importance as a direct cause of minor offences of dishonesty" (p. 52). Not only in Sweden and England, but also in Ireland, Gibraltar, and Norway there was no evidence that crime and gambling are related.[24] Nor in the U.S., where the myth of gamblers being criminals seems to originate, is there any evidence that the American bettor was involved in criminal acts, other than the placing of the illegal bet itself.[25]

Why, then, do people associate gamblers with criminals, with unstable family lives, with irresponsible family behavior?

Recall that in the model two classes of people had greater incentives to participate in games of chance: the poorer and those who suddenly became significantly poorer. The first group planned participation, while the second decided suddenly to participate.[26] However, the motivation to commit crimes could not be linked at all to the first group—only to the second. But if the first group represents the majority of the buyers, while only a small fraction of the second gambles on criminal acts (the rest gamble on an entrepreneurial one), it should not be expected that in general gambling and crime will be strongly correlated. Only some who have suddenly lost a significant fraction of their wealth may decide to gamble both on games of chance and on criminal acts—and they may represent only a small fraction which cannot justify a legal ban on gambling.[27] Indeed, the facts suggest that only a tiny fraction did both gamble and commit crimes. However, the histories of people belonging to this category have been the most interesting ones to capture the attention of social scientists and writers and may have led to both misleading interpretation of and negative attitudes toward gambling in general. But it may be useful to note that opinion polls reveal that the relatively poor favor gambling and do not condemn gamblers; it is only those who are richer who condemn them.[28]

The negative attitude toward gambling in the U.S. may stem from a different reason: namely, that when it was *outlawed,* the games were supplied by the "underworld." But it is a mistake to form opinions on the relationship between crime and gambling where gambling is outlawed, since such opinions become just matters of semantics.

If gambling does not seem to have any of the negative consequences attributed to it why should it not be legalized? Recall the familiar arguments. Fear of the destabilizing effects of compulsive gamblers has been one argument. Yet, as the data show, compulsive gamblers are a tiny fraction of the gambling population (the majority of those who gamble do so in order to become richer). Their existence cannot be used as an argument against gambling: there are compulsive eaters, compulsive drinkers, compulsive workers, compulsive watchers of TV. Yet eating, drinking, working "too much," and watching "too much" TV are activities that are not forbidden— yet. The second popular argument has been that since gambling leads to crime, laws must be passed to restrict it. First, as shown, the facts are that this relationship holds for a tiny number of people only. Second, the existence of a correlation in the U.S. is *not* a proof that gambling "causes" crime. Rather, the very fact that gambling was outlawed created the incentive for the underworld to capture this sector. The Swedish experience in which gambling was outlawed until 1930 shows the losses that such legal restrictions may impose on the society: the Swedish loss was incurred not only because resources were allocated for prisons and police in order to enforce the anti-gambling legislation, but also because the Swedes gambled on the English soccer games, thereby smuggling out substantial amounts of Swedish currency.[29] Once gambling was legalized, the criminal elements that were involved with the smuggling and the gambling disappeared.

In conclusion, the model has predicted that the relatively poor plan to gamble regularly because through this channel they hope to change their relative position in the distribution of wealth significantly.[30] The objections that have been put forward to restrict gambling are prejudiced, and, if adopted, may only do harm directly (closing an avenue for becoming richer) or indirectly (since prosecutions and prisons cost money). Clemens France, in "The Gambling Impulse," advocated gambling with this statement: "Man can live without pleasure, but not without hope." Forbidding gambling closes one channel of hope, and who can tell what the indirect costs of such an act will be? What other channels of hope will people invent instead?

2. On Social Instability and Progressive Taxation

To make benevolence scientific is the great problem of the present age.

Arnold Toynbee, 1884

The second prediction of the model concerned the motivation for participating in criminal acts against property. It has been shown that one is more likely to start committing crimes when one's position in the distribution of wealth has suddenly been significantly worsened. Thus we can say only that when people become relatively poorer they become more likely to commit criminal acts, that is, acts not in accordance with existing laws or customs. The criminal act is done with the expectation that, if undetected, it may make it possible for one to improve one's position in the distribution of wealth. A related prediction was that if the position of many people is worsened, they may gamble on revolutionary ideas that advocate a redistribution of wealth.

Numerous empirical analyses for the United Kingdom, the United States, and Canada have shown that when more people have fallen below a "poverty line" the crime rate has increased.[31] Consistently, such changes have proven to be good explanatory variables for crime rates, independent of the specific econometric method used, or the other explanatory variables introduced in the statistical tests. Of all the implications of the model these two are probably the least surprising and have in fact been suggested since ancient times: (1) that an unexpected worsening of position and the perception of inequality lead people to commit crimes; and (2) when a relatively large group's position in the distribution of wealth is worsened, the probability of revolutions increases.[32]

Assuming that this relationship among crime, revolutions, and changes in the distribution of wealth is valid (even if it has not been precisely defined within a formal model), one should expect to find that societies have gambled on several methods in order to restore their stability in case of significant changes in the distribution of wealth. The model suggests three such methods: when more people become poorer, more can be spent on police force (which would increase the probability of being caught), punishments can become harsher (which would diminish the benefit from crimes), or payments can be transferred to those who have become poorer (which would diminish the incentives to commit crimes).

The difference among the three methods is that the last seems to be the only one that can successfully deal with the negative effects of unexpected changes in the distribution of wealth. There are several reasons for this: first, we do not always know when expectations change, and when a significant group in the society perceives that its position in the distribution of wealth has been worsened. Thus an increase in instability cannot be *prevented* by the first two methods.[33] Second, the strategies of changing either the police force or the structure of punishments require lengthy legislative actions, and may thus be unsuccessful in preventing the increase in crime rates. In contrast, transfer payments to those who have become poorer provide an automatic mechanism that can prevent both the increase in crime rates and the gamble on a revolutionary act, although how much these transfer payments should be we cannot say.[34] The next sections both examine the issues formally and discuss evidence on the emergence of such forms of taxation.

Social Instability and Progressive Taxation: Formal Analysis

He who considers things in their first growth and origin, whether the state or anything else, will obtain the clearest view of them.

Aristotle, *Politics*

In order to justify the existence of a progressive wealth tax, we will use the theory presented in chapter 1. We will show that since within this model the actors know who is more likely to commit crimes or to gamble on revolutionary ideas, it is easy to understand by what methods these acts can be prevented. It is shown that progressive taxation provides such a method; how other social scientists have justified progressive taxation is discussed in the next section.

Recall the conditions that lead one to commit a crime (the proof for the revolutionary gamble is similar): let H be the amount one expects to gain by committing a crime, a burglary for instance. Suppose there is a probability, p, of being caught, in which case the burglar will bear a punishment of monetary value, C, C being either a fine or the monetary value of the time in prison. If one commits the burglary and is caught, one's utility becomes $U(W_0 - C, \alpha(W > W_0 - C) \mid \alpha(W > W_0))$, while if one is not caught one's utility becomes $U(W_0 + H, \alpha(W > W_0 + H) \mid \alpha(W > W_0))$, given that $\alpha(W > W_0)$ represents one's expected position in the distribution of wealth. An individual will not commit the crime if the expected utility of committing it is less than his utility if he does not. In mathematical symbols this statement can be written as follows (recall the definition of the probability, p, in appendix 1.2):

$$(1) \qquad pU(W_0 - C, \alpha(W > W_0 - C) \mid \alpha(\,\cdot\,)) + (1 - p)\, U(W_0 + H,$$
$$\alpha(W > W_0 + H) \mid \alpha(\,\cdot\,)) \mid U(W_0, \alpha(W > W_0) \mid \alpha(\,\cdot\,))$$

However, when the distribution of wealth changes, people whose relative position has worsened will tend to commit more crimes, and the crime rate will increase. The intuitive explanation for this result is clear: when *relative* wealth diminishes, the probability that one will commit the same crime that one was previously reluctant to commit increases because one is suddenly "outdone" by his fellows. Such an act enables one to have a chance to restore one's position in the distribution of wealth to its expected level.

Consider an economy in which people's position in the distribution of wealth changes. The probability that crimes are committed increases. The richer people who are the victims of these crimes are aware of this (people's expectations of crime rates are derived from the model presented in chapter 1). The richer people in our model are also aware that crime is one method of wealth redistribution, that maintaining people's position in the distribution-of-wealth stable will diminish the crime rate, and that crime rates can also be reduced by increasing expenditures on self-protection (locks, guards, doormen, etc.) or on police and crime enforcement. The last methods increase the probability of being caught, p, and would lower the crime rate. At the same time these methods diminish the wealth of the relatively rich: methods of self-protection do so directly, while expenditures on police and crime enforcement do so indirectly, due to higher taxes.

The richer people can thus choose among several strategies for diminishing crime rates: (a) they can increase expenditures on crime prevention and raise p, the probability that the criminal will be arrested and kept in jail; and (b) they can increase transfers to those who are poorer either by charities or by the government (i.e., by paying more taxes). Both methods represent transaction costs paid by the relatively rich with the expectations of increasing the stability of the society and, indirectly, increasing their wealth. In conclusion, if people behave according to the view presented here, then a progressive tax on wealth can be viewed as the means of maintaining social stability.

Let us prove these statements formally: let Π_0 and Π_1 be, respectively, the probabilities of a rich person being the victim of a crime before and after the introduction of a progressive taxation scheme and W_0 and W_1, respectively, this rich person's expected wealth when he is not the victim of a crime, before and after the introduction of progressive taxation. Let H_0 and H_1 be, respectively, the average expected costs due to crime before and after the introduction of progressive taxation. The following relations must hold according to our theory:

(2) $\Pi_1 < \Pi_0$; $W_1 < W_0$; $H_1 < H_0$.

Then, for the relatively rich to prefer the introduction of a progressive taxation scheme the following condition must hold (when for simplicity's sake we omit the conditional statements):

(3) $\Pi_0 U(W_0 - H_0, \alpha(W > W_0 - H_0)) + (1 - \Pi_0)$

$$U(W_0, \alpha(W > W_0)) < \Pi_1 U(W_1 - H_1, \alpha(W > W_1 - H_1))$$

$$+ (1 - \Pi_1) U(W_1, \alpha(W > W_1))$$

If Π_1 and H_1 are sufficiently diminished by the progressive tax scheme, this condition can be fulfilled *even* if W_1 is smaller than W_0. Then, both the rich and the poor are better off if a progressive-taxation scheme is introduced. This conclusion is reached *without* making either an interpersonal comparison of utilities *or* assuming a diminishing marginal utility of wealth, comparisons and assumptions that social scientists have made in order to justify a progressive tax—more about their views below.

However, this argument, while justifying a progressive-taxation scheme, does not say anything as to just *how* progressive the scheme should be. This depends on the responsiveness of Π (the probability of a rich man being a crime victim) to changes in the tax rates, on the effect the redistribution of wealth has on both crime rates *and* the total wealth, and on how responsive *p,* the probability of detection, is to an increase in police expenditures.

How Have Others Justified Progressive Taxation?

One guess that has been made to explain the existence of progressive taxation is that the marginal utility of wealth is diminishing (for an excellent, critical summary of this view see Blum and Kalven 1953); a dollar is "obviously" worth less to a richer than to a poorer man.[35] Suppose, therefore, that a poor man and a rich man have the same utility function of wealth.[36]

(4) $U = U(W)$

W being the wealth. Suppose also that the marginal utility of wealth, $U' = dU/dW$, is positive, and that the marginal utility of wealth is diminishing. Then, if there are two men, one rich with a wealth of W_r and one poor with a wealth of W_p ($W_p < W_r$), it follows that

(5) $U(W_r) > U(W_p)$ and $U'(W_r) < U'(W_p)$

i.e., while the total utility is greater for the richer man than for the poorer one, the marginal utility of the last dollar is smaller for the rich man. Thus, in order to ask a sacrifice from a richer man equal to that asked of a poorer man *in terms of utility,* we need to tax the *last* dollar of income of the rich more than the last dollar of income of the poor.

This guess has been criticized for many reasons. The first and more trivial criticism was that this justification of progressive taxation relies on an interpersonal comparison of utilities which in general is not made in economic analysis. This is a criticism leveled at all welfare economics, mitigated here by the fact that since the rich man and the poor man are assumed to have the same utility function, comparison between utilities makes some sense. The second, more important criticism is that the behavioral hypothesis of decreasing the marginal utility of wealth cannot really be inferred from the traditional models of individual behavior in which risk does not exist. In these models if one utility function with decreasing marginal utility of wealth is consistent with the preferences of the consumer, any monotonic increasing transformation of this function is also consistent with them. But the monotonic transformations that transform the marginal utility of wealth from a diminishing into an increasing function always exist.[37] Thus the decreasing marginal utility of wealth cannot be inferred from these models. It may be useful to note that already in the late thirties Henry Simmons asserted that all the theorizing "explaining" progressive taxation summarized above was "intellectual rubbish" (we use his words), and that the case for progression was no more and no less than the case for mitigating "unlovely" economic inequality.[38] As can be inferred we agree with his statement: what we have tried to provide here was a basis for a *positive* analysis of "unlovely" inequality.

The second strand of justification for a progressive taxation scheme has been based on the guess that people are "risk-averse." This guess was made when some economists (for a summary of their views see Friedman and Savage 1948 and Arrow 1970) have observed that individuals choose to insure themselves against risks, and that it has been empirically remarked that the premium, H, they pay for most insurances is higher than the actuarially fair ("fair" being defined by the following mathematical condition: if the individual may lose an amount L with probability $p,$ then the "fair" premium is one that equals the expected cost, i.e., pL. But it has been found that the premium is greater than pL). This behavior seemed consistent with the assumption of risk-aversion, an assumption that implies that an individual always prefers certain wealth to a gamble in which he can either lose or win an amount h with a probability of $\frac{1}{2}$. In turn, this preference implies that the marginal utility of wealth must be decreasing. Thus, if everybody were risk-averse, one could justify a progressive income-tax schedule.

The problems with this reasoning are the same ones as raised by making the guess that people are in general "risk-averse." This guess, while it "explains" people's desire to insure (in a nonverifiable way), forbids them to gamble on "unfair" bets (bets like lotteries, in which the expected loss is greater than the expected gain) at the same time. But our day-to-day observation shows that people both insure themselves and gamble. Friedman and Savage (1948) tried to correct this inconsistency between the guess and its predictions by assuming that at a very low level the marginal utility of wealth decreases, in a middle region it increases, while in a region of high wealth it again decreases. Although this pattern would indeed be consistent with the fact that people both gamble and insure themselves, the justification for a progressive tax would be lost. Moreover, this restriction on the shape of the utility function has led to predictions inconsistent with a new set of evidence: as both Alchian (1953) and Markowitz (1952) have pointed out, one of the results of Friedman and Savage's view would be that the relatively rich would neither insure themselves against events in which big losses occur with small probabilities nor gamble on unfair bets. But this is what they do. Thus one cannot justify progressive taxation by assuming that everybody is "risk-averse."[39]

Let us now compare our view of progressive taxation with some recent studies that have appeared in the economic literature. Hochman and Rogers (1969) did show that income redistribution could be chosen voluntarily by the rich as an "efficient" move. Their guess was that the richer people's utility function depends on the poorer people's wealth—this guess, which suggests that the rich are altruists, cannot be verified. In contrast, the guess presented here does not presume altruism. Rather, it suggests that the relatively rich realize that a progressive wealth tax provides insurance for maintaining the somehow achieved social order.

It must be pointed out, however, that the contrast between our view and that of Hochman and Rogers is not as sharp as this last sentence would imply. Suppose that at one point the richer people did gamble on a progressive wealth tax as a method for stabilizing the society. Later generations, who grew up within a society where such a tax system already existed, would view it as a custom and not inquire into its origin. Then an outsider who did not examine the emergence of this system could look at it and view it as a sign of altruism—this is how Hochman and Rogers's views can be interpreted. Thurow (1971) makes another guess, namely, that some "social-welfare function" does exist. But this guess prevents him, as well as all "welfare" economists, from making a testable prediction. Let us reemphasize the difference between their methodology and ours: we do not assume that people are altruistic, that there are social-welfare functions, or that a comparison can be made between people's utilities. Our guess is that people's behavior is shaped by their relative wealth and unexpected changes in it, a guess that, as shown, can be examined in numerous ways[40]—one examining the emergence of progressive taxation is presented next.

Progressive taxation, which has been introduced by various governments since the end of the nineteenth century, was originated by Bismarck[41] (although it should be noted that redistributions from the richer to the poorer enforced by customs have characterized primitive societies).[42] First the tax was imposed on inheritance; only later, on incomes. While many writers have mentioned that a belief in "justice" led to its imposition, when one looks at the facts one gets a clearer perception: the taxes were introduced in order to prevent social unrest not only in the form of increased criminal acts, but also in the form of gambling on revolutionary ideas.[43] The observations of several writers are useful at this point. John Stuart Mill (1848) noticed that when preindustrial methods of social protection became obsolete, pressures for new forms of protection would rise: the laboring classes in an industrial society would no longer accept authority and protection from the old ruling classes and show obedience to them. That, of course, does not mean that they would not accept them from the *new* classes once the appropriate ideology (someone's "lucky hit") had been found. Rimlinger's (1982) observations support this view:

> Social security was perceived by labor leaders as a serious threat to the solidarity of the labor movement, which was of course also Bismarck's view . . . The hypothesis suggested here is that so long as a labor movement is insecure and fighting for its survival, it cannot be counted on to press for policies that might appear to make the government the chief protector of the workers. Revolutionary labor movements find it particularly difficult to support measures that might strengthen the

existing order by promoting stability and by weakening the workers' revolutionary ardor. [p. 155]

Indeed, as Stern (1977), Pinson (1954), and Craig (1978) have documented, Bismarck expressed his views quite clearly. He recommended a wide variety of social legislation to lure workers from potential and actual socialist loyalties, hoping to repress subversion and remove its causes. At the end of 1878 he remarked that "if the worker had no more cause for complaint, then the roots of socialism would be chained off." Stern (1977) concludes that

> Bismarck's aim was unambiguously political: in the grim moments of dependency, the lower classes should come to know the state as a source of succor and should not have to rely on indigent family, on indifferent employer, or on a socialist party. St. Vallier recognized the magnitude of Bismarck's program: "[it] is more comprehensive, more audacious, more dangerous than the others; he wants to fight the socialists by borrowing their aims and by making the state the pivot of all workers' organizations." [p. 219]

But why were progressive taxation as well as other welfare systems introduced toward the end of the nineteenth century? As argued in my previous book, when population increases and becomes more mobile and customs break down, governments will start substituting for roles that kin have played in redistributing wealth. Yet this substitution takes time: one must first invent the ideas that represent a political validation of the new claims for redistribution, and then figure out how to implement them. Thus at first the ideas may be vague and slogans may be followed. Yet people will be willing to follow such slogans, and their acts will reflect the sentiments articulated here (is there an alternative?). As mentioned above, Bismarck's social-insurance program represents the classic case of a redistributive policy being implemented as a response to threats: Bismarck believed that increased insecurity induced people to gamble on the revolutionary doctrines of the Social Democrats. As Rimlinger (1982) notes, when there were no such pressures (as in France before World War I) neither the idea that the working class was a victim of exploitation nor the threat to social stability posed by revolutionary syndicalism had much impact (this is not surprising, as the discussion in the previous chapter, on Machiavelli's views in particular, has already suggested).

Rimlinger's view raises an additional major question: why was the legislation introduced first in Germany and only later elsewhere? While a pattern, a regularity, can be explained, uniqueness cannot be. One can only guess. The guess in this case refers to the individual (Bismarck in this case) who shapes the course of events since people bet on his ideas. Such individuals play a central, yet random, unpredictable role, and their appearance and ideas cannot be explained with precision. One may examine, as Rimlinger and Pinson do, however, the different attitudes among nations, attitudes shaped by their whole history, and attribute to them, in part, the tendency to bet on some ideas. Pinson, for example, notes that the industrialization of Germany brought with it all the signs of instability associated with such a change. According to him the program of state social legislation met with relatively less resistance in Germany than elsewhere since the tradition of state paternalism was stronger, going back to William I, William II, and Frederick II, the latter calling himself "the king of the beggars." But even Pinson, in explaining the emergence of the various forms of social legislation, seems to put the major emphasis on Bismarck's personality, shaped by both the German tradition and the Caesarism of Napoleon III in France and the Tory democracy of Disraeli in England.

In conclusion, a methodological point in our approach to the issue of progressive taxation: a custom *cannot* be explained by its current utility, but only becomes intelligible through its past history—this has been the line of argument pursued until now,[44] and it will be pursued further in the rest of the chapters. The clue to this line of research can be found in Aristotle's *Politics*, as the epigraph to one of the sections in this chapter suggests.

3. On Creativity; or, Why Do Productivity and Profits Decline?

The model presented in the previous chapter made one further prediction in addition to those on gambling and crime, namely, a prediction regarding people's creativity. This prediction will be examined below.[45] However, first we shall see how the terms "profits" and "productivity" must be redefined within the model and can be related to "creativity."

Entrepreneurs, Productivity, Profits: Definitions

An entrepreneur is an individual who gambles on novel ideas.[46] The entrepreneurs we remember are those who made lucky hits. This view of the entrepreneurial act means something strikingly different from the view that some individuals take risks because they are "risk-lovers" and has far-reaching consequences for understanding how firms operate.

A "risk-lover" is defined in traditional economic analysis as an individual who is willing to participate in an unfair gamble. This trait is attributed to abstract conditions related to the shape of the utility function. In contrast, in the model one's willingness to take risks depends on the *change* in one's position in the distribution of wealth. When one's wealth diminishes one is more likely to gamble on an idea on which one was previously reluctant to gamble—in business, management, technology, or the arts.

This view of the entrepreneurial act enables one to perceive the difference between the definitions of "risk" and of "uncertainty." When people's position in the distribution of wealth is stable, they still take risks: the orange they buy may be rotten, the car they buy may turn out to be a lemon, and they may lose $500 in a bet. Yet, ex post, the involvement in these acts may lead to no significant changes in their position (their effects have been discounted). Therefore, within this model, their willingness to undertake entrepreneurial acts (either criminal or noncriminal) is not altered; they just react in a customary way. Such states can be characterized as "risky." In contrast, when one's position in the distribution of wealth is significantly altered, one is more likely to undertake an act one has never undertaken before. Such a state can be defined as "uncertain" since it leads people to gamble on new ideas that did not come to life before (that had only existed latently in their minds).[47]

This distinction enables one to understand the meaning of "profits" in the model. When only risky activities are undertaken in the economy, the outcomes can lead only to a redistribution of wealth. Thus, if one could properly measure profits, they would be equal to zero in the aggregate: one individual's gain is another one's loss since wealth is only redistributed but not created. In contrast, the gambling on novel ideas involves, by definition, the creation of new "wealth" when a lucky hit is made. Thus an outsider would then measure positive economic profits.[48]

But since people are more likely to gamble on novel ideas when their relative wealth diminishes, the resulting profits cannot be attributed to "greediness" or be interpreted as a sign of increased welfare. For it is one's diminished welfare to start with that led one to gamble on novel ideas, the outcome of which is only measured by *others* as positive profits. However, the individual who made the profit will view it as a return on his willingness to bear a risk in the new circumstances in which he suddenly found himself.

If these views on profits are correct, the meaning of "profit maximization" becomes very different from the one used in the traditional theory of the firm. In that theory, production functions and relative prices are all given, and the only thing "profit maximizers" are expected to do is to adapt to changes in relative prices, but never to change them by themselves. But according to our view the way individuals can make profits is either by gambling on novel ideas and making lucky hits[49] *or,* if one understands the human attitudes we have outlined here (even if one does not know how to articulate them), by giving proper incentives to employees to gamble more frequently on novel ideas. An employee who does so could be defined as having "managerial skills."

How can good managers increase the probability that their subordinates will not carry out their job mechanically but will gamble on novel ideas and increase their effort and productivity? Our views suggest several methods that can be used to achieve such a goal. One is a specific wage structure: namely, having significant differences in the wages of similarly trained individuals and having fewer positions at the higher level. Such a structure would provide strong incentives for some of the lower-paid individuals to gamble on novel ideas. At one point or another in their life cycle their expectations are frustrated. Assume that one expected to achieve a higher-paid job at the age of forty with a probability p. One's expected position in the distribution of wealth was calculated by taking into account this probability. Suppose, however, that at the age of thirty, or thirty-five one realizes that the probability of getting the job at the age of forty (or ever) has diminished. What is one to do? If our views are correct, the probability that one will make greater efforts, gamble on new ideas, and try to "prove" oneself increases. Thus a pyramidal structure provides, in the model, the incentives to induce people to make greater efforts and thus become more creative and productive.[50]

So there is some truth in the Xerox advertising that says, "Productivity is: a manager who helps subordinates grow" (this view contrasts with the one provided by standard economic analysis in which productivity is just an unexplained residual). But such advertising gives only part of the truth. As already shown, the diminished position may lead not only to greater probability that people will gamble on innovations, but also to "criminal" acts like the gossip on Cunningham-Agee.[51] In a *Wall Street Journal* article (15 July 1982, p. 9), entitled "Pitting Workers against Each Other Often Backfires, Firms Are Finding," the following episodes were described: to induce branch managers to perform better, a European bank encouraged them to compete against each other to produce the best results. The winners were promised bonuses. The outcome was that the bank discovered that a manager had steered customers to rival banks rather than help another branch manager win the bonus. Similar events took place in a Los Angeles insurance company where the management ranked offices according to how frequently they distributed disability payments on time. What happened was that when one office got a claim that was meant for another, workers frequently used the mail rather than the telephone to redirect the information in attempts to lower the performance of competitors. These examples show how difficult the task of "profit maximization" is—when the management tries to raise productivity and profits by the methods suggested by this model, it walks on the razor's edge.

Thus the question is whether there are methods other than significant differences in wages that can increase people's productivity. The model suggests several such methods: cancelling conditions involving job security, tenure, or automatic promotions. But we wish to make it clear that we do not suggest that it is always "good" to do these things, or that it is always "good" to promote the entrepreneurial trait. For it is a trait that results from people's unhappiness. It is only important to realize what choices a society faces and when, if the majority so decides, it is likely to take a chance.[52]

Before presenting some evidence on creativity, it is useful to summarize the picture of firms that emerges from the model. New firms emerge because there are people who either by chance or unexpected fluctuations in the distribution of wealth became entrepreneurs and can organize production (i.e., sell an idea) better than others: the firm is their channel for becoming richer. Profits can be the result of chance, lucky hits (call them innovations or discoveries), or managerial skills, which in this model are defined as the skills that enable one to build a structure of rewards within the firm that promotes productivity. The practical implication of this view is that companies should realize that profits start from ideas, and *not* from "research budgets," that good management is as important as the good skills of workers (since such management can provide greater motivations for workers and translate ideas to "products"), and that "money" or "higher wages" do not produce innovation (a theory of how decision makers within firms behave is presented in further detail in appendix 2.2).

This view of the behavior of firms and of entrepreneurs, while strikingly different from that which exists today in the economic literature, bears similarities with Schumpeter's (1927) view: that it is not so much "money" that prevents one from carrying out an entrepreneurial act as an ordered way of life:[53]

> Actually among the obstacles in the way of the rise of an industrial family, eventual lack of capital is the least. If it is otherwise in good condition, the family will find that in normal times capital is virtually thrust upon it. Indeed, one may say, with Marshall, that the size of an enterprise . . . tends to adapt itself to the ability of the entrepreneur. [pp. 157–58]

> But in considering this process of expansion, we come upon [another] reason for the varying success of business dynasties. Such expansion is not simply a matter of saving and efficient routine work. What it implies is precisely departure from routine. . . . Most members of the class are handicapped in this respect. They can follow suit only when someone else has already demonstrated success in practice. Such success requires a capacity for making decisions and the vision to evaluate forcefully the elements in a given situation that are relevant to the achievement of success, while ignoring all others. The rarity of such qualifications explains why competition does not function immediately even when there are no outward barriers . . . and this circumstance . . . explains the size of the profits that often eventuate from such success. This is the typical pattern by which industrial fortunes were made in the nineteenth century, and by which they are made even today. [pp. 158–59]

Are profits due to "exploitation" within my model? The answer is NO.[54] It is true that in this model individuals will not receive the "value of their marginal productivity" (the concept cannot even be defined). But they should not: they may achieve their productivity because they work in a team where entrepreneurs and managers have found the incentives to make them more productive. In the absence of such management the marginal value of their productivity would be lower (although it is impossible in this case to speak about one's marginal productivity, since this depends on other people's ideas—more about this point in appendix 2.2).

In the next section we present the statistical relationship between the number of patents applied for and registered and measures suggesting changes in the distribution of wealth. Other directions in which the predictions on creativity have been examined are presented in Brenner 1983b and in the chapters that follow.

Evidence on Patents

The model has predicted that those whose relative position has suddenly, significantly worsened are likelier to engage in different sorts of gambles which may help them regain their expected position in the distribution of wealth. One of the possible gambles is to undertake either a research or an inventive activity which, if successful, may result in a patentable invention. Therefore, the model predicts that when such disorder occurs the number of patents may increase (notice that within the model it is meaningless to speak about their "value").

We must nevertheless mention the many drawbacks in the empirical test on patents we shall make. First, the number of patents, as an indicator of inventive activity and creative thinking, is an extremely downward-biased measure of "novel ideas." For not all innovations are patentable, and though innovations in literature and arts are "patentable" by copyright arrangements, these, however, are very difficult to interpret: nobody would argue that textbooks are innovative. Examples of the fact that not all innovations are patentable are many: a new administrative procedure streamlining organization is as innovative as any new patented technology. (As Drucker wrote in a recent *Wall Street Journal* article: installment selling may have had a greater impact on economics and markets than most of the "great advances in technology" in this century.)[55] The invention of "la nouvelle cuisine" may be as productive as the invention of any new, patented, electronic gadget. But these innovations are not patentable. New art forms or music constitute an even greater problem, since it is impossible to patent such ideas perfectly and enforce property rights (even experts have difficulties at times, distinguishing Braque from Picasso). The Beatles' innovation in music was so productive for England that it significantly improved its balance of payments—yet their contribution does not appear in any statistical measure of "innovation." And how could one measure their originality—or Picasso's? How many of Picasso's ideas could be traced to Cézanne, Braque, and primitive art? Thus patents will always be a downward-biased measure of the innovative activities taking place in the economy. Yet, if the fraction of new ideas that is patentable stays stable, the tests should confirm the relationship between the number of patents and changes in people's position in the distribution of wealth.

In spite of the fact that we are aware of these difficulties, we have tried to see whether or not patents applied for and registered (in the U.S. and Canada) depend on changes in the distribution of wealth. Patents are one of the few statistical measures available which may give some idea on the rate of gambling on novel ideas in these societies. The data includes patents applied for and granted to both individuals and corporations. Including corporations in the data set does not constitute a drawback, since as shown in the previous section, their hierarchical structure insures that researchers will be subject to the same incentives because of inequality as individuals. Their creativity, too, is due to their fluctuating position (because of altering expectations) in the hierarchical structure, and measures of inequality, if accurate, should capture such fluctuations.

The U.S. Study on Patents

The data on patents have been adjusted for population growth. Separate statistical analyses have been made for the annual number of patents applied for and registered. The reason for looking at both measures is that the model suggests, in fact, that the number of *applications* should be taken into account. For some people may have been unaware of ideas that others have been working on, and many could have come up simultaneously (more or less) with similar original ideas. While no patents will be registered, such applications should be taken into account for the purpose of verifying the implications of the model. Also, some people may have decided to commit a "criminal" act and make an application by imitating or stealing ideas, hoping that they may not be caught. Since within the model the motivation to commit such criminal acts is the same as that which induces others to gamble on entrepreneurial ones, it is likely that the statistical analyses, where the number of such "original ideas" is included in the dependent variable, would provide better results (as indeed it does). The reason for looking at the second measure—registered patents—is that one would like to have an idea as to whether or not after eliminating the aforementioned "criminal" cases the model's predictions are still not refuted. It turns out that they are not.

In order to test our prediction that patents per 100,000 inhabitants will increase whenever the distribution of wealth significantly changes, we need a measure of this changing situation. Since we wanted to test our theory over as long a time period as possible, we had to choose the following measure: the percentage of U.S. wealth in the hands of the 5% of the population who are the wealthiest. This measure, computed by Kuznets (1953) for the years 1919–46 is the only one available. For the period 1947–78 with which Kuznets did not deal, we have computed this same percentage from Bureau of Census publications. This measure, too, is a very imperfect one, and in Brenner (1983b), where the implications on crimes were tested, the percentage of families below one-half of the median income was used, a more accurate reflection of a change in people's position in the distribution of wealth. But since this measure could only have been computed for a relatively short period of time only, we chose the longer, Kuznets-type series.

Since this measure of inequality is so imperfect we have decided to introduce several other variables which could provide improved information on what happens to the distribution of wealth in the economy over time.

The existence of government income-maintenance transfer programs either to supplement income when it is judged inadequate (such as food stamps, welfare payments, medicare, and so forth), or to soften the impact of unemployment (by unemployment insurance) have an effect on the distribution of wealth in the economy, an effect that will not be reflected by changes in the number of people in the top 5%. The more extensive these programs are, the less the same sudden decrease in the revenues of some groups will be taken as a signal of diminished relative wealth. For the government is now expected to supplement one's income. The existence of such programs constitutes a disincentive to inventive activity. We tried to measure the influence of these programs by using the real 1958 dollar value of the social-welfare expenditures per 100,000 inhabitants as an explanatory variable. We expect that the greater these payments, the less patents will be granted. Let us reemphasize why we use this variable: if the measure of change in people's position in the distribution of wealth was appropriate, we would not have used it since the measure would already incorporate its effects. But our measure is most unlikely to capture it. Thus this variable is an indicator of whether or not people expect their position to be significantly altered.

Another variable likely to influence inventive activity, though it might not be captured by either of the two previous measures of inequality, is the unemployment rate. When the increased unemployment is expected to be of a more permanent duration, the distribution of wealth *is* expected to change, and the number of patents is likely to increase. However, if the increased unemployment rate is expected to be only temporary, its effects on the number of patents might be insignificant. If the first perception is correct, neither our measure of inequality nor the measure of government transfers will adequately capture the effects of these altered expectations on inequality, but the unemployment rate may provide the additional source of information. In this case one would expect that an increase in unemployment rates would have a positive effect on patents, although one cannot say whether its effect will be significant or not (it turns out that except for one test among eight the effect of the unemployment rate is insignificant).

An additional variable is included to complement all the previous measures of perceived disorder, namely, a proxy for the weakened customs and family ties. The reason for including this explanatory variable is the following: one of the traditional ways in which people have insured themselves against unexpected drops in their wealth, when population was smaller and less mobile, was the institution of the extended family. Lending without taking interest, but assuming the responsibility for caring for the elderly, the sick, and the unlucky, were some of the facets this insurance took. The trend of diminishing family size is a continuous one in the U.S. (and in what is called "Western civilization" in general): when unexpected, such a trend leads to fluctuations in one's relative position in the distribution of wealth—these issues as well as those defining how "wealth" depends on family ties and customs are discussed in detail in Brenner (1983b). The effects such changes have on the distribution of wealth are very unlikely to be captured by conventional measures of inequality—for several reasons: one's perception of what constitutes a "family" is different from the legal definition, and one's expectations of whether insurances protected only by customs will be fulfilled or not are hard to quantify.[56]

However, the weakening of family ties is positively correlated with urbanization rates. Thus we have decided to use the data on these rates to serve as a proxy for the effects the unexpectedly weakened family ties could have had on the distribution of wealth. We have chosen this variable rather than one reflecting the change in "family" size because of the many definitional problems related to the latter.

The list of variables used and their sources are given in table 2.6. The

Table 2.6   List of Variables Used in the U.S. Regression Analysis

Variable name	
PAT[1]	Number of patents granted to U.S. residents per 100,000
PAT[2]	Number of patents applied for per 100,000
U	U.S. percentage of civilian labor force unemployed
INE	Percentage of U.S. wealth detained by the wealthier 5% of the population
URB	Urbanization rate measured by the fraction of the population living in cities
PUB	Social-welfare expenditures under public program per 100,000 in 1958 constant dollars
WAR	Dummy variable equal to 1 for war years 1941–45 and 0 otherwise

Sources: PAT[1]: Office of Technological Assessment Forecast Report, 7th report.
PAT[2], U, URB: 1919–76: *Historical Statistics of the U.S.*
1970–78: *Statistical Abstracts of the U.S.*
INE: 1919–46: Williamson and Lindert (1980), from Kuznets (1953).
1947–78: Computed from *Statistical Abstract* of the U.S.
PUB: Same as U; deflated by CPI.

added variable, WAR, is a dummy variable equal to 1 for the years 1941–45 of World War II when the draft took potential inventors away and the inventive capacity of the economy was geared to the war demand. We have used 0 for other years. A priori one cannot say what effect the WAR variable may have had on patents: on one hand, potential inventors were drafted; on the other hand the defense effort must have spurred inventions.

At first a least-square regression model of the form

(6) $PAT_t^i = \beta_0 + \beta_1^i \, INE_t + \beta_2^i \, U_t + \beta_3^i \, URB_t + \beta_4^i \, PUB_t +$

$$+ \beta_5^i \, WAR_t + \epsilon_t \qquad i = 1, 2$$

has been estimated, where $i = 1$ indicates the analysis for patents granted, and $i = 2$ for patents applied for. The variables are defined in table 2.6 in which t refers to the observations for year t, ϵ_t is the residual for year t, and all variables (except WAR) are logarithmic transformations of the ones defined in table 2.6. $\beta_1, \beta_2, \beta_3$ were expected to be positive (but β_2 insignificant), and β_4 negative. The estimations obtained were:

$PAT^1 = \quad 3.8 \quad + \quad 4.4 \quad INE \quad + \quad 0.2 \; U \quad + \quad 1.8 \quad URB \; -$
$\qquad\qquad (1.4) \qquad (4.3) \qquad\qquad (0.4) \qquad\quad (2.1)$

$\qquad\qquad\qquad\qquad\qquad\qquad\qquad 0.01 \;\; PUB \; - \qquad 0.01 \;\; WAR$
$\qquad\qquad\qquad\qquad\qquad\qquad\quad (-0.5) \qquad\qquad (-0.24)$

$R^2 = 0.6$

$D.W. = 0.8$

$F(5,54) = 16.5$

$PAT^2 = \quad 5.5 \quad + \quad 2.9 \quad INE \; - \qquad 1.08 \;\; U \; + \; 0.85 \quad URB \; -$
$\qquad\qquad (16.3) \qquad (4.3) \qquad\qquad (-2.2) \qquad (1.55)$

$\qquad\qquad\qquad\qquad\qquad\qquad\qquad 0.04 \;\; PUB \; - \qquad 0.23 \;\; WAR$
$\qquad\qquad\qquad\qquad\qquad\qquad\quad (-2.4) \qquad\qquad (-4.5)$

$R^2 = 0.77$

$D.W. = 0.7$

$F(5,54) = 37.8$

The numbers in parentheses below each coefficient are the $t-$statistics. All the coefficients (except the unemployment rate in the second case) have the predicted signs, but only the measure of inequality is statistically significant in both.

However, since the Durbin-Watson statistics suggest autocorrelation, we have corrected for this and obtained:

$PAT^1 = \quad 4.6 \quad + \quad 3.3 \quad INE \; + \; 0.2 \;\; U \; + \; 0.7 \quad URB \; -$
$\qquad\qquad (5.7) \qquad (2.3) \qquad\qquad (0.3) \qquad\quad (0.5)$

$\qquad\qquad\qquad\qquad\qquad\qquad\qquad 0.02 \;\; PUB \; - \qquad 10.06 \;\; WAR$
$\qquad\qquad\qquad\qquad\qquad\qquad\quad (-0.5) \qquad\qquad (-0.7)$

$R^2 = 0.3$

$$\text{PAT}^2 = \underset{(10.6)}{5.3} + \underset{(2.4)}{2.2} \text{ INE} - \underset{(-0.2)}{0.1} \text{ U} + \underset{(1.7)}{1.4} \text{ URB} -$$

$$\underset{(-3.4)}{0.07} \text{ PUB} - \underset{(-3.0)}{0.1} \text{ WAR}$$

$R^2 = 0.5$

The results seem better: all the coefficients that are statistically significant have the expected signs.

Let us now examine how the statistical results vary when we use explanatory variables that may be better proxies for the variables appearing in the model. Within the model, unexpected changes in people's position in the distribution of wealth influence the number of applications for patents. But this reaction may be due not necessarily to observed changes, but to *expected* ones. It may thus be argued that the measure of realized positions in a year may not be a good proxy for expected changes in such positions in the distribution of wealth. Therefore two alternative measures are used in the following tests: First, the lagged measure of inequality, INE_{t-1}, has been used, the estimated model being:

$$(7) \qquad \text{PAT}_t^i = \gamma_o^i + \gamma_1^i \text{ INE}_{t-1} + \gamma_2^i \text{ U}_t + \gamma_3^i \text{ URB}_t + \gamma_4^i \text{ PUB}_t +$$

$$+ \gamma_5^i \text{ WAR} + \epsilon_t^i$$

all the rest of the variables being the same as in (6). Again, we expected that the coefficients γ_1^i, γ_2^i, and γ_3^i will be positive, and γ_4^i and γ_5^i negative. Since the ordinary least-square estimations suggested autocorrelation we corrected for it, and the results became:

$$\text{PAT}^1 = \underset{(6.2)}{4.6} + \underset{(2.5)}{3.2} \text{ INE} + \underset{(0.6)}{0.4} \text{ U} + \underset{(0.5)}{0.7} \text{ URB} -$$

$$- \underset{(-0.4)}{0.01} \text{ PUB} + \underset{(0.8)}{0.07} \text{ WAR}$$

$R^2 = 0.3$

$$\text{PAT}^2 = \underset{(11.9)}{5.4} + \underset{(2.7)}{2.2} \text{ INE} - \underset{(-0.1)}{0.01} \text{ U} + \underset{(1.6)}{1.2} \text{ URB} -$$

$$- \underset{(-3.2)}{0.07} \text{ PUB} - \underset{(-2.9)}{0.15} \text{ WAR}$$

$R^2 = 0.6$

All the statistically significant coefficients have the expected sign: the inequality measure is significant at the 5% level in both tests, the public-transfers coefficient is statistically significant only when the dependent variable represents applications for patents, while the unemployment rates are insignificant in both cases.

The second assumption on the formation of expectations was that inequality in any year, t, was forecasted from several of its past values. We used the ARIMA estimation to obtain the forecast of inequality in year t from its past values, and then used the forecasted value, FOINE, to estimate this model:

$$(8) \qquad \text{PAT}_t^i = \gamma_o^i + \gamma_1^i \text{ FOINE}_t + \gamma_2^i \text{ U}_t + \gamma_3^i \text{ URB}_t + \gamma_4^i \text{ PUB}_t +$$

$$+ \gamma_5^i \text{ WAR} + \epsilon_t^i$$

Since the first estimation suggested autocorrelation, we corrected for it and obtained:

$$\text{PAT}^1 = \underset{(3.7)}{3.8} + \underset{(2.4)}{0.9} \text{ FOINE} + \underset{(0.7)}{0.6} \text{ U} + \underset{(0.9)}{1.3} \text{ URB} -$$

$$- \underset{(-0.8)}{0.03} \text{ PUB} + \underset{(.07)}{0.06} \text{ WAR}$$

$$R^2 = 0.28$$

$$\text{PAT}^2 = \underset{(9.0)}{5.6} + \underset{(1.4)}{0.3} \text{ FOINE} + \underset{(.02)}{0.1} \text{ U} + \underset{(1.2)}{1.09} \text{ URB} -$$

$$- \underset{(-3.4)}{0.08} \text{ PUB} + \underset{(2.9)}{0.16} \text{ WAR}$$

$$R^2 = 0.48$$

Again, all the coefficients have the expected sign, but only the measure of inequality is significant in both examinations.

Miraculously, the results seem quite consistent with our arguments. We say "miraculously" because the data were so inaccurate that we really did not expect very good results. With very bad data, it would indeed be suspicious if the results we obtained were too good. The fact that all the coefficients are in the direction that we have expected (although not all of them are significant) was a pleasant surprise. One may ask why these results have been presented here if we have so little confidence in the data? The reason is mainly methodological: the exercise shows one more way in which the predictions on creativity derived from the view of human nature presented here can be tested. Other predictions on the reaction of minorities who have been discriminated against have been examined in Brenner (1983b). Although it is useful to look at the diagrammatical representations of the statistical analyses in appendix 2.1, they suggest that the value of these statistical games may not be entirely methodological after all.

The Canadian Study on Patents[57]

In Canada, data on the distribution of wealth have not existed for as long a period as in the U.S. This has limited the time period under consideration to 1951–79. In order to have more observations we have thus tried to do a time-series and cross-sectional analysis relating a measure of inequality in each of the five Canadian regions to the number of patents granted to residents of these regions (we could not find data on the number of applications).

In this study the dependent variable is again the number of patents per 100,000 inhabitants granted to Canadian residents per region.

In order to test our model, we must relate this variable to a coefficient of the relative bettering or worsening of people's situations in the five *regions* considered. We have chosen two measures: the percentage of families whose income is inferior to one-half of the median family regional income and the percentage of individuals of at least 14 years old whose income is inferior to one-half of the median personal regional income (as argued above, one would expect that the former could provide a better measure of changes in the distribution of wealth). The drawbacks of these measures were that they could only be computed since 1951 (when Statistics Canada published its first survey of regional income distribution), and that only since 1971 have such data been computed on an annual basis. Thus, for the twelve years between 1951 and 1970 that are missing, the inequality measures have been estimated by linear interpolations.

One further drawback with these inequality measures was that until 1971 they were figured for gross, pretax income only and thus did not take into account the redistributive effects of progressive taxation (although they did consider income transfers); we can thus expect that these measures are biased upward.

As in the U.S. study, other explanatory variables have been introduced as an attempt to compensate for some of the missing factors that affect fluctuations in the distribution of wealth. These variables are the regional urbanization and unemployment rates, but not the size of the public income-maintenance programs, since their effects are included in the measure. The list of variables and their sources are given in table 2.7.

We estimated the following models:

(9) $\quad \mathrm{PAT}_t^i = \gamma_1^F + \gamma_1^F \mathrm{INF}_t^i + \gamma_2^F \mathrm{U}_t^i + \gamma_3^F \mathrm{URB}_t^i + E_t^i$

$\quad\quad \mathrm{PAT}_t^i = \gamma_0^I + \gamma_1^I \mathrm{INI}_t^i + \gamma_2^I \mathrm{U}_t^i + \gamma_3^I \mathrm{URB}_t^i + E_t^i$

where the index t refers to the year t, i to the region i, F to the equations using the family inequality measure, and I to the one using the individual inequality measure. The variables are all logarithmic transformations of the ones defined in table 2.7 (where these variables were rates, we took the logarithm of 1 plus the variable).

Table 2.7 List of Variables Used in the Canadian Regression Analysis

Variable name	
PAT	Number of patents granted to Canadian residents per 100,000
U	Percentage of active population unemployed
INF	Percentage of family units whose income is inferior to one-half the median family income
INI	Percentage of individuals age 14 and over whose revenues are inferior to one-half the median personal income
URB	Urbanization rate measured by the percentage of the population living in cities

Sources: PAT: 1950–65: Report of the Patents Overseer

1966: Report of the General Registrar of Canada

1967–71: Annual Report—Consommations and Corporations Canada
1974–79:

1972–73: Patents Bureau Gazette

U: Statistics Canada bulletins nos. 11–505F, 71–201

INF, INI: Statistics Canada bulletins nos. 13–503, 13–504, 13–512, 13–517, 13–521, 13–528, 13–534, 13–544, 13–207

URB: Statistics Canada bulletin nos. 92–608, 92–807; also 1956 census data

We expect γ_1^F, γ_1^I, γ_2^F, γ_2^I, γ_3^F, and γ_3^I to be all positive (although γ_2 is not necessarily significant). The estimates of (9) were:

$$PAT = -\ \ 3.2\ +\ 2.3\ INF\ -\ \ \ \ 12.4\ U\ +\ 9.2\ URB$$
$$(-7.6)\ \ \ \ (1.5)\ \ \ \ \ \ \ \ \ \ \ \ \ (-9.9)\ \ \ \ \ \ \ (21.2)$$

$$R^2 = 0.8\ \ \ \ \ \ \ \ \ \ D.W. = 0.43\ \ \ \ \ \ F(3,141) = 197.6$$

$$PAT = -\ \ 2.6\ -\ \ \ 0.3\ INI\ -\ \ \ \ 12.4\ U\ +\ 9.2\ URB$$
$$(-4.4)\ \ \ (-0.1)\ \ \ \ \ \ \ \ \ \ (-9.4)\ \ \ \ \ \ \ (20.5)$$

$$R^2 = 0.8\ \ \ \ \ \ \ \ \ \ D.W. = 0.42\ \ \ \ \ \ F(3,141) = 194$$

The inequality of the family coefficient has the expected sign but it is not significant, while the coefficient of the individual inequality is neither positive nor significant (but we did expect that the latter measure of inequality would perform less well than the former). In both cases the coefficient of the urbanization variables is positive as expected. On the other hand the coefficient of unemployment is negative. However, where corrections have been made by various methods for both autocorrelation and heteroskedasticity (see appendix 2.1), the only significant change was that the unemployment rate turned out to be insignificant, as expected.

In conclusion, it seems a surprise that with the very bad data available the coefficients all came out in the expected direction, even if not all of them are significant.

Conclusion

One may now raise this question: How can one distinguish between our views on innovations, productivity, and creativity and alternative ones that have been given? Fortunately, we can give one answer to this question: for the moment there seem to be *no* alternative views that explain why people bet on novel ideas. In economics there are models of research and development (R & D), but they do not define what these words mean, and have nothing to say on people's creativity. These models seem to suggest that "money" produces innovation. Why one would say that, we do not know: it is not true that those born rich are more creative, although they certainly are able to spend more on "R & D."

Thus our answer to the question is this: since, for the time being, our hypothesis seems to be the only one on "creativity" that can be contradicted, we do not know what predictions of which alternative theories we should compare them with.[58] Until now the evidence presented both in this book and in Brenner (1983b) does not seem to contradict our views.

Appendix 2.1:
Evidence on Patents:
Diagrams and Comments

The Canadian data-set pools time-series and cross-sectional observations. Thus the usual assumptions of homoskedasticity and independence of errors of the ordinary least-square model are unwarranted. In order to correct for this drawback we assumed that for each year the error is the sum of three shocks: one that is truly random, one that is region-specific, and one that is year-specific. Then we estimated the model using the Fuller-Battese (1974) method. The results obtained were:

$$PAT = 0.02 + 1.5 \ INF - 2.4 \ U + 2.2 \ URB$$
$$(0.04) \quad (1.) \qquad (-1.8) \quad (2.9)$$

$$PAT = 0.02 + 0.8 \ INI - 2.3 \ U + 2.4 \ URB$$
$$(0.04) \quad (0.45) \qquad (-1.7) \quad (3.1)$$

While the sign of the unemployment and urbanization coefficients are the same as previously, the sign of inequality becomes positive, although not significant, in both estimations. Other methods yielded similar results.

Figure A2.1 shows what is hiding behind the statistical results: one line represents the number of patents, while the other measures lagged inequality. A glimpse makes it clear that the lines representing patents and measures of monetary inequality move in the same direction between 1919 and 1947, but seem totally unrelated after that. As explained in the text there might be several reasons for this peculiarity: after 1946 government transfer programs started to play an increasing role in shaping expectations for people's position in the distribution of wealth which have not been captured by customary measures of inequality. Also, the last thirty years have seen drastic changes in family structure which, again, while significantly changing the distribution of wealth, have not been captured by the customary measures of inequality. However, the effect of these changes is captured in the statistical analysis by the various demographic variables. In spite of its limitations, the diagram may be suggestive for those skeptical of statistical analyses, for whom the way in which the results of such analyses are stated in the chapter provides little insight.

Fig. A2.1 Patents Registered in the United States and Lagged Inequality

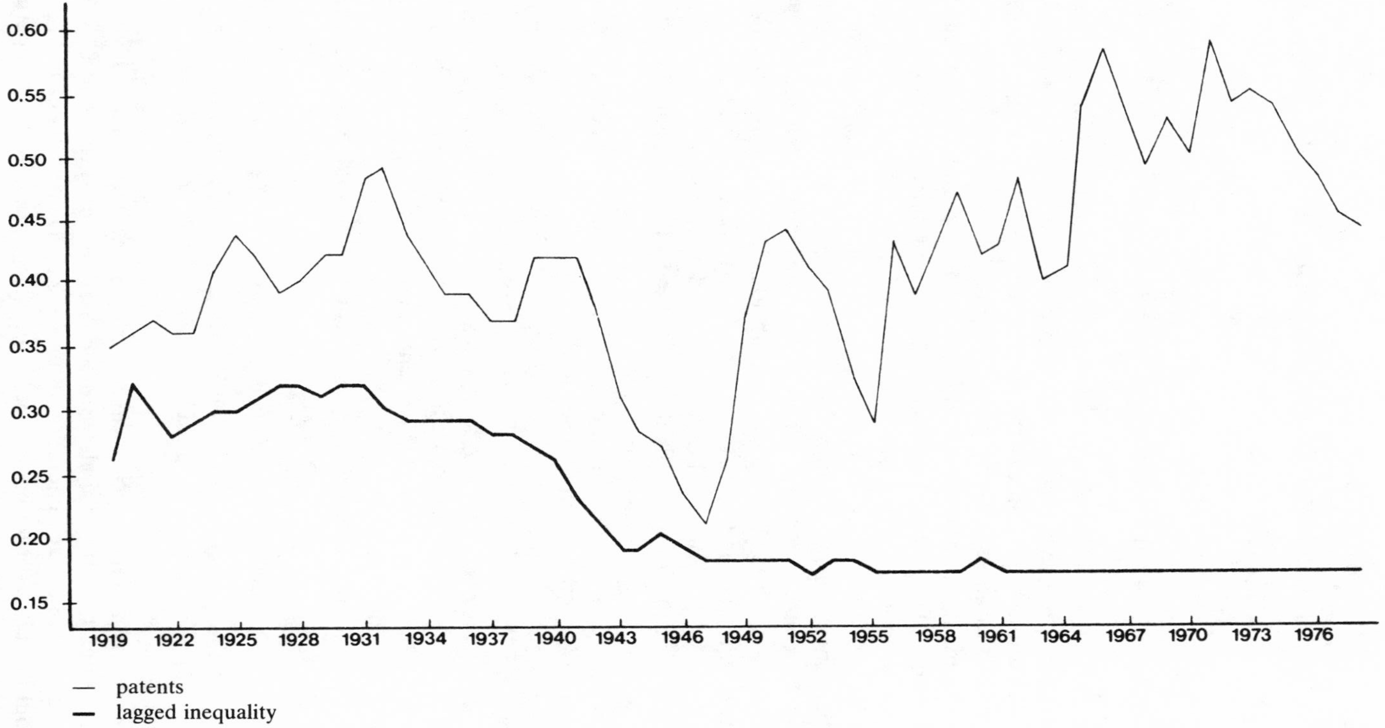

— patents
▮ lagged inequality

Appendix 2.2:
Firms and International Trade:
An Alternative Viewpoint

Chapter 1 has presented one aspect of international relationships—conflict—while the last section in this chapter presented an alternative view of the behavior of firms, a subject associated with trade. This appendix combines the two arguments in order to shed light on some aspects of international trade and at the same time elaborate the theory of the firm.

To a large extent the theory of international trade still represents a reexamination of inherited doctrines based mainly on Smith, Ricardo, Marshall, Heckscher, and Ohlin, that is, on refinements of the neoclassical theory, in which "free trade" always seems beneficial.

However, both in the past and recently some economists have advanced the noncustomary opinion that the predictions of the inherited doctrines were inconsistent with the facts. Two major inconsistencies that have troubled economists (Williams already in 1929, and more recently Klein 1977) were the so-called Leontief, or trade, paradox and the growth paradox (discussed by Williams, but reemphasized by Abramovitz in 1956). The first section of this study briefly discusses these paradoxes and the explanations that have been suggested to resolve them. There now seems to be unanimity that the distinguishing feature of foreign trade has been the exchange of *recently* invented technologies in the U.S. for raw materials and products that were invented a while ago (i.e., the U.S. was exchanging products of entrepreneurs for products whose production became customary). But what model can make the prediction that one country will have the relative advantage of coming up with new ideas and implementing them (i.e., with the trait economists define as "entrepreneurship")?

The model presented in chapter 1, together with an elaboration of the arguments presented in the last section of this chapter, will suggest my answer.

Theories and Facts

Some thirty years ago economists tried to explain two paradoxes of economic performance in the U.S.: the "growth" paradox and the "trade" paradox.[59]

In terms of the neoclassical approach, growth depends on savings and the accumulation of "capital." But the facts are that growth in U.S. has been mainly due to the generation of new ideas. These facts have been pointed out both by Abramovitz (1956), who showed that during the period 1900–1950 increases in "capital" and "labor" inputs played only minor roles, and by Solow (1957) whose recalculations for the same period suggested that only ⅛ of the per capita growth of the U.S. GNP is to be attributed to a per capita increase in capital stock and that about ⅞ can be attributed to a "pure" increase in productivity. If this is the case one must explain "growth" by explaining why productivity increases. But many economists perceive productivity gains as exogenous variables, i.e., as being determined "outside the economic system."[60] This opinion is similar to saying that innovations are like "manna from heaven," a viewpoint that is not, in fact, an "explanation."

The "trade" paradox is linked to the previous one. According to classical trade theory, the U.S. should have been exporting "capital"-intensive and importing "labor"-intensive products. However, the facts are that the U.S. exported both types. In fact, the distinguishing feature of foreign trade was the exchange of newly invented and implemented technologies in the U.S. for already existing, perfected products. Leontief (1956) has pointed out the inconsistency between the predictions of the neoclassical model and the facts, and it was Vernon (1966) who made the observation on the existing pattern of trade. In another study Gruber, Mehta and Vernon (1967) underscored Vernon's views:

> From capital and labor cost considerations, therefore, attention has turned to questions of innovation, of scale, of leads and lags. Approaches of this sort have tended to stress the possibility that the United States may base its strength in the export of manufactured goods upon monopoly advantages, stemming in the first instance out of a strong propensity to develop new products or new cost-saving processes . . . [Another line of argument] takes off from the observation that entrepreneurs in the United States are surrounded by a structure of domestic demand for producer and consumer goods that is in some respects a forerunner of what will later be found in other countries. [p. 17]

Notice that the explanations rely on the concept of "entrepreneurship", "new products" and of being "forerunners"; Gruber Mehta, and Vernon (1967) conclude:

> In sum, one derives a picture of high research effort being correlated with industries that experience substantial trade surpluses. These research-intensive industries, although large and concentrated, are not systematically capital intensive. It is in these industries that the U.S. trade advantage lies. [p. 28]

Or, as Gruber and Vernon put it in the 1968 book, the U.S. has been trading "brain for brawn" (p. 266)—my model, as elaborated in section 2 of this chapter gives precision to this statement.

One attempt to explain both the growth and the trade paradoxes was made by Denison (1967), who suggested that the U.S. has an educational advantage. He estimated that about half of the productivity advance can be attributed to investment in human capital, as measured by increased years of schooling. According to this approach the traditional distinction between "labor" and "capital" was simply inappropriate.[61] However, some economists have been skeptical of using the human-capital approach to shed light on the two paradoxes (for a summary of such views see Klein 1977). First, it has been argued that it is unclear whether the increase in the years of formal schooling is a cause or an effect. Although it can be assumed that more schooling increases productivity and helps to generate new ideas, it can also be assumed that productivity gains and new ideas are due to exogenous reasons and that more years of schooling just represent more leisure and more entertainment, which become available *because* of the increased productivity.

Let us assume, however, that more years of schooling do increase productivity and generate new ideas. If this view is accurate, one should be able to observe a continuous, positive relationship between the two variables. But, not surprisingly, as my views suggest, no correlation between new ideas and formal education seems to exist either on the "micro" or "macro" (historical) levels. Beaton, for example, found that self-employed businessmen in the U.S. had average earnings of $6,200 more than salaried employees after adjustment for formal education and so-called ability (measured by Beaton on an IQ scale), while Klein (1977), Ohmae (1982), and Peters and Waterman (1982) point out that innovations have rarely been made by Ph.D.'s, in some industries even by engineers, or by people with highly specialized training. They discuss numerous cases: the automobile industry in both the U.S. and in Japan, which was pioneered by people who had little formal education, and the large number of industries in which the innovations were made by newcomers. If one takes a historical perspective, Denison's arguments become even less convincing: formal education in the U.K. rose continuously after the Industrial Revolution, but productivity did not. Also, if formal education was the major explanatory variable, how can one explain the significant divergence in growth rates among countries with similar levels of formal education (a question recently raised by Olson 1978, 1982)? After all, one should note that formal schooling provides information on our ancestors' knowledge, that it transfers these memories (stored in books and computers) to the younger generation, and that more formal education (and the higher wages being paid for it, on average) only implies that the people receiving it are expected to have memorized more. But how does formal schooling *generate* new ideas? Or, does it?

But according to the model presented in chapter 1 and its appendixes, if in the U.S. customs are weaker than elsewhere (for historical reasons), and institutions to protect people's position in the distribution of wealth have not yet evolved, the facts on U.S. "productivity" and international trade patterns that have been observed will seem consistent with the model's predictions.[62] The U.S. is expected to have more entrepreneurs at every level of the decision-making process.

Before relating my views of human nature to the views of some economists, it seems useful to devote the next section to the theory of the firm as perceived in the light of the model. This analysis is needed: (a) to shed further light on and to give more precision to the words "entrepreneurship" and "productivity"; (b) to present a theory of firms in the light of this model. After all, the decision to carry out international trade is made, in part, at the level of the decision makers within firms; (c) to point out more precisely why some societies have a relative advantage in generating new ideas, while others have an advantage in perfecting existing products.

2. Entrepreneur, Productivity, and a Theory of the Firm: An Alternative Viewpoint

What do "firms" do? In order to answer this question let us clarify what the word "firm" means. The words "firm" and "corporation" are defined by the law: the law defines their rights and obligations. In what sense, one might ask, can one speak about their "behavior"?

Suppose that a family owns a restaurant and that only family members run it. If one attributes predictable behavior to this firm, it means that one attributes predictable behavior to whomever is making the final decisions. Thus, if one assumes that families behave in a particular way, one must also assume that family firms behave in exactly the same way. In other words, the motivations attributed to the household and the family-owned business must be the same. Yet the economic literature (with few exceptions)[63] assumes that households and firms have different motivations. Why this is so is not always clear.

Why do particular households decide to cooperate and sell something to other people? *To put it simply, some people may believe that they can become richer by using this channel since they have an idea that they expect to sell.* The idea may be a novel one, or just an imitation of one that has worked. Such ideas may come to life when a number of individuals cooperate directly, each making a contribution: one person may be the entrepreneur who offers the idea, the other may provide the managerial ability, while the third may gamble on the venture and provide the money. Such organizations may eventually become relatively large, and the owners (stockholders, for example) may no longer have direct control over those who make decisions within the corporation. How can the behavior of such a corporation be characterized? Again, the answer is that the "behavior" of such a corporation reflects the behavior of the individuals who make the decisions within it. Thus, in order to understand either the behavior of such corporations, or that of family-owned businesses, one may try to reduce the problem to that of understanding how people make decisions (i.e., how they make up their minds). This view has been presented in chapter 1. Let us show now how this view enables us to sharpen the definitions of additional words associated with the behavior of "firms" that have not yet been discussed (the words "risk," "uncertainty," "profits," "entrepreneurial acts" have been already defined in Brenner (1983b) and within this chapter).

If the views of entrepreneurial acts and profits defined by this model are accurate, the meanings of "profits" and of "profit maximization" become totally different from the ones used in formal theories of the firm. In the neoclassical view, "production functions" and relative prices are all given to "firms," and the only thing that "firms," or "profit maximizers" do is adapt to changes in relative prices. "Firms" play a totally passive role. In contrast, according to my view "profits" are associated both with unexpected changes in the distribution of wealth and the resulting increased gambling on novel ideas of people whose relative position has been worsened. In addition to gambling on novel ideas in technology or the arts, some may also gamble on new ideas as to how to reorganize production by providing incentives to employees to gamble more frequently on novel ideas. According to the model one can achieve this goal if one understands the motivations of people as outlined here (even if one may not know how to articulate them); an individual who does so can be defined as having "managerial skills": chance and these entrepreneurial skills determine the size of firms within this model (just as in Marshall's and Schumpeter's views, quoted in the text).

How can managers increase the probability that their subordinates will increase their efforts and gamble more frequently on new ideas (acts that can be viewed as leading to increased "productivity")? The model suggests several methods that can achieve such goals. One is to recommend a specific wage structure: that is, to set significant differences in the wages of similarly trained individuals and have fewer positions at the higher levels. Such a structure provides incentives for some of the lower-paid individuals to gamble on novel ideas. For at one point in their life cycle their expectations are bound to be frustrated. Assume that there is some probability that one will manage to get a higher-paying job (or a job with greater status) at the age of thirty or forty. One's aspirations took this probability into account. But suppose that one suddenly realizes that the chances of fulfilling one's expectations have significantly diminished. What will such an individual do? This individual will now make greater efforts and gamble more frequently on new ideas. Thus a hierarchical structure provides the right incentives to induce people to increase their productivity. Other methods by which managers can increase their employees' productivity have been discussed in the text.

But the scenario described above already shows precisely what "team-work" is in the model, and why it induces the contractual form called "the firm." Some managers may find the proper incentives for some employees to become more productive. However, these incentives may be good for some people and not for others. Thus people's productivity may increase under the schemes and strategies of some managers but not under those of others (as an intuitive example consider how differently orchestras perform under different conductors). In this sense one can speak about the productivity of a team, and see the motivation for the contractual form that characterizes firms.

How will the ideas that people gamble on differ in the two types of societies defined earlier? Suppose that while in both societies some new ideas can be patented, in one customs are stronger and they, too, serve to enforce property rights. New ideas and implementations that can be patented must be rather drastically different from already patented ideas and methods of production; if they are "too close," or if words cannot accurately capture the distinction, they will not and cannot be patented (can a new managerial strategy, the idea of the "credit card," for example, not to mention new musical and artistic languages, be patented?). Perfecting the mechanism of a car, polishing the final product, advertising a product with a new slogan, reflect ideas that can be protected by customs (by guaranteeing lifetime employment in a company, for example), but not always by laws. People will gamble more frequently on ideas involving the polishing and perfecting of existing products in a society in which customs are expected to be enforced than in a society in which there are no such expectations and only the returns on "big" ideas are expected to be protected (in such a case "smaller ideas" are expected to be stolen and imitated with impunity). This prediction, together with the one derived in the previous section on the motivation to become an "entrepreneur," can be applied to patterns of international trade: in one society individuals will gamble more frequently on "big," new ideas, on implementing and exporting them, while in the other, if in spite of all insurances the entrepreneurial trait surfaces, individuals will gamble more frequently on ideas for perfecting existing products (imported ones in particular), and eventually export their craftsmanship.

In conclusion, the model and the vocabulary used in this model not only shed light on the so-called paradoxes of growth and trade, but also present a theory of firms in terms of the individuals who make decisions within them. It has been shown that while one society has an advantage in inducing the trait defined as "entrepreneurship," another has an advantage in promoting craftsmanship. The words "entrepreneurship" and "craftsmanship" suggest that the search for answers requires looking at the motivations of individuals, within some well-defined social and legal structure (an approach attempted by Klein 1977 and suggested, in part, by Gruber and Vernon's already-mentioned discussions).

4. Similarities and Differences in Alternative Approaches: Concluding Comments

As noted in the first section, several studies have concluded that international trade must be explained in terms of "entrepreneurship," "managerial decisions," and "productivity gains," terms that refer to the analysis of firms. This has been the approach taken here. Thus the comparison in this section will be made not with traditional "international-trade theories" (which anyway shed no light on the facts and just use undefined words like "labor" and "capital"), but with alternative theories of decision making and of firms. Before doing that, it may be useful to reemphasize that my model's terminology is not really new—it just makes more precise the words used by Gruber et al. when describing patterns of international trade: namely, that "brain" has been exchanged for "brawn."

The similarities between the model of the firm presented here and the views of Simon (1959) have already been pointed out in detail in appendix 1.1, where his view of human nature, of entrepreneurial acts in particular, has been quoted.[64] The views of other economists with regard to the entrepreneurial act have been presented in Brenner (1983b). So let us turn now to the other words associated with the theory of the firm. While some economists have argued that no distinction should be made between "uncertainty" and "risks," others assert that such a distinction should be made, although they do not succeed in defining how this should be done. Knight was one who saw a relationship between "profits" and uncertainty and argued that only uncertainty rather than risks is related to "pure profits." For all the other economists (in the neoclassical tradition), "profits" are associated with "non-contractual costs."[65] These amounts are equal to the difference between total receipts and total contractual costs, and are received by the owner of the "entrepreneurial capacity." Friedman (1962) clarifies this definition:

> This term [profits] is, however, somewhat misleading. The actual non-contractual costs can never be determined in advance. They can be known only after the event and may be affected by all sorts of random or accidental occurrences, mistakes on the part of the firm, and so on. It is therefore important to distinguish between actual non-contractual costs and expected non-contractual costs. The difference between actual and expected non-contractual costs constitutes "profits" or "pure profits"—an unanticipated residual arising from uncertainty. Expected non-contractual costs, on the other hand, are to be regarded as a "rent" or "quasi-rent" to entrepreneurial capacity. They are to be regarded as the motivating force behind the firm's decisions. [p. 99]

Several comments should be made on this quotation. First, Friedman uses the word "uncertainty," in spite of the fact that elsewhere he argued that no distinction should be made between this term and the word "risks." Second, the occurrence of "profits," according to this view, is purely accidental. Third, Friedman views "entrepreneurial capacity" related to "rents," although as the next comment shows he was well aware of the fact that when defining the first term he was only playing with words:

. . . implicit in the view we are adopting [is] the notion that each individual can, as a formal matter, be regarded as owning two types of resources: (1) His resources viewed exclusively as "hired" resources— what his resources would be if he were not to form his own firm . . . (2) A resource that reflects the difference between the productivity of his resources viewed solely as hired resources and their productivity when owned by his firm—we may call this "Mr. X's entrepreneurial capacity" . . . It should be emphasized that this distinction between the two types of resources is purely formal. Giving names to our ignorance may be useful; it does not dispel the ignorance. A really satisfactory theory would do more than say there must be something other than hired resources; it would say what the essential character-istics of the "something other" are. [p. 95]

The model and arguments presented in the first chapter provide exactly such a theory and suggest getting rid of the vocabulary which only led to the illusion that our ignorance has been dispelled. On this issue Fritz Machlup's (1946) views of the firm should also be mentioned. He (as well as Buchanan [1979]) stresses the subjective interpretation of the variables manipulated by "firms":

It should hardly be necessary to mention that all the relevant magni-tudes involved—costs, revenue, profits—are subjective—that is, per-ceived or fancied by the men whose decisions or actions are to be explained . . . rather than "objective" . . . Marginal analysis of the firm should not be understood to imply anything but subjective esti-mates, guesses and hunches. [pp. 521–22]

While this view is consistent with mine, the rest of Machlup's suggestion, that whatever managers choose to do can be called "profit maximization," is not. According to my view, managers just gamble and see "profits" as something radically different from what this term means for Machlup (and in the "marginal analysis").

Many economists have made the observation that firms do not maximize profits and that new insights can be gained by looking at the motivations of the individuals who make decisions: Becker's (1957) *Economics of Discrim-ination*, Simon's (1957, 1959) *Models of Man* and his articles on decision making, Alchian and Kessel's (1962) article, Baumol's (1977) book all make this observation. Other economists have asserted that the behavior of the firm can be best understood by looking at the motivations of its managers. This is the approach suggested by Williamson's (1964) *The Economics of Discretionary Behaviour: Managerial Objective in a Theory of the Firm* and by Marris's (1964) *The Economic Theory of Managerial Capitalism*. Both examine the behavior of firms by looking at the motivations of their man-agers; both assume that the manager's goal is to do the best for *himself*—in this sense my point of departure and theirs are the same. However, neither Williamson nor Marris provides a verifiable theory which can explain how managers make decisions when facing risks, a theory provided in this book—this is the core of the difference between our approaches.

As to the roles of uncertainty and teams in explaining the existence of firms, Coase (1937) argued that an examination of the behavior of firms must start by asking what leads to their emergence. Coase suggests that the existence of "transaction costs," i.e., the costs of contracting all resources through "markets," can provide such an explanation. However, what these "transaction costs" are precisely, he does not define. According to the model, one's entrepreneurial talent, the interaction between managers and employees in particular, that leads to the provision of greater efforts and the more frequent gambling on ideas, is the characteristic feature of the "firm" that cannot be provided by an alternative arrangement. Among economists, only Coase (1974) has also suggested that one should examine the "market for ideas" in general. Yet nowhere in his study does he suggest that examining this issue means discussing how people *think*, and that if one had such a model, all aspects of human behavior should be examined within it. For isn't "thinking" what human behavior is all about?

Alchian and Demsetz (1972) suggested that this is indeed what a "team process" or a "firm" is all about when they wrote:

> The problem of economic organization, the economical means of metering productivity and rewards, is not confronted directly in the classical analysis of production and distribution. Instead, that analysis tends to assume . . . zero cost . . . means, as if productivity automatically created its reward. We conjecture the direction of causation is the reverse—the specific system of rewarding which is relied upon stimulates a particular productivity response. If the economic organization meters poorly, with rewards and productivity only loosely correlated, then productivity will be smaller; but if the economic organization meters well productivity will be greater. [p. 779]

Yet Alchian and Demsetz neither give precision to their views, derive verifiable predictions from them, nor relate their explanation to uncertainty and people's attitudes toward risks.

In addition to Simon's earlier-mentioned views, the model also bears relationship to those of both Schumpeter and Veblen: both have emphasized that the perception of risk and of risk-taking depends on features of the society one lives in. Schumpeter (1942), for example, argued that if redistributive policies are expected to take place, the supply of entrepreneurs will diminish—the similarities and differences between his and my views are discussed elsewhere.[66] Veblen (1933) argued that the success of technological entrepreneurs is followed by a decline, since financial entrepreneurs who build on the ideas of others only try to insure themselves by preserving the organizations built around the lucky hits of the technological entrepreneurs.

Finally, some views from the business-marketing literature should be mentioned. In contrast to those expressed in the standard economic literature, the views expressed by executives on the decision-making process within firms (as summarized in Solman and Friedman 1982 and Peters and Waterman 1982) seem similar to mine, and they are even stated in a similar language.[67] Here are some typical statements:

> "When it came right down to it" says Moore [once an IBM financial executive], "I was betting on people. That's what you're always really doing . . . You can quantify all you want—and believe me, people do. But you can never replace the need for a judgement call" . . . [and another executive said] "Yes . . . After all the quantification, I think acquisitions are made on the basis of vision, not numbers." [Solman and Friedman 1982, p. 115]

> We don't kill ideas, but we *do* deflect them. We bet on people . . . You invariably have to kill a program at least once before it succeeds. That's how you get down to the fanatics, those are really emotionally committed to finding a way—any way—to make it work. [Peters and Waterman 1982, p. 230; italics in original]

The executive's statement in the second quotation suggests that the model of human behavior in terms of which he was thinking seems similar to the one presented here.[68] Also notice that the words "vision," "judgment calls," "betting on people" (the latter phrase appearing in both quotations), have precise meanings in the theory of the firm presented here, but hardly any in the neoclassical models.

In conclusion, the model of human behavior suggested here seems to shed light on both existing patterns of international trade (patterns that seem inconsistent with the predictions of existing theories) and activities associated with firms—and *without* creating an artificial, new language.[69]

Appendix 2.3:
On the Methodology of a Uniform Approach

*After a while I began to understand the issues and realized, to my
amazement, that arbitrators were not really concerned about who
was right and who was wrong, what was true and what false, but each
was looking for twists and turns to justify his party and to contradict
the arguments of his opponent.*
                          Isaac Bashevis Singer, *In My Father's Court*

The model presented in this book and in my previous one suggests a uniform
paradigm for the social sciences: it incorporates many of the basic ideas that
have been offered through the ages as points of departure for examining
human behavior. Sociologists and Marxists alike have shared the perception
that classes, inequality, and reference groups are central to understanding
human behavior (though there are many differences with their approaches,
as is shown both in the last chapter of this book and in Brenner 1983b). The
model looks first at the motivations of individuals. But in contrast to the
neoclassical economic approach "stability" matters rather than "efficiency"
(a word that cannot even be defined within my model). People's reactions
depend on what happens to others, and governments fulfill obvious roles.
Decision makers within governments try to maintain stability and, when
population fluctuates and customs break down, they gamble on new policies
in order to substitute for roles previously played by family, kin, and customs. There are no "invisible hands" within my model. Nor does one need
Adam Smith's assumption of the existence of the "instinct to truck and
barter." "Trucking and bartering" just represent some of the channels for
becoming richer; but there are others as well. To some extent the views of
the psychologists also seem to be vindicated: after all, emotions seem to be
the creative fuel for thinking, and "stress" (interpreted here as fear of being
hindered by either internal or external opposition, or competition) leads
people to take entrepreneurial gambles in art, crime, business, science, and
politics. Perhaps psychologists should be reminded of what Freud wrote
about himself and view his theories in this light: "I am not really a man of
science, not an observer, not an experimenter, and not a thinker. I am
nothing but by temperament a *conquistador*—an adventurer, if you want to
translate the word—with the curiosity, the boldness and the tenacity that
belong to that type of being" (quoted in Jones 1961, p. xi; italics in original).
   The model also shows that once unexpected shifts in people's position in
the distribution of wealth are taken into account, there is room for both

leadership and chance in understanding events. But it is *impossible* to appraise their influence with precision. While one can, perhaps, accept the idea that if not Columbus then somebody else would have discovered America, can one believe that the features of World War II would have been the same without Hitler? Of course, as the model suggests, the appearance of leaders (i.e., of political entrepreneurs) and of entrepreneurs in general depends on perceived changes in the distribution of wealth. But their ideas depend on probabilities, and thus one can never explain the individual case.

Finally, some methodological points: since people don't behave according to categories, there are no categories in my model. The terms "economic," "sociological," and "psychological" represent mere inventions of new vocabularies. The motivations are either to get richer or to maintain stability, and there is no need to invent technical words to describe them. In contrast to other approaches in the social sciences, mine looks at the human condition through spectacles that show how people make up their minds. Neither the economic, nor the sociological, nor the political-science approach has much to say on this subject. Last, but not least: if the arguments in this book seem simple, that may be, after all, a misleading perception. They are complex. Only the *vocabulary* I use is simplified relative to that currently used by most social scientists and most philosophers: the model makes this simplification possible.

The reader has probably already perceived from this and the previous book that much attention is paid to linguistics. Since my view is that unclear language is a sign of unclear thinking, I have tried first to define a few words with precision and then to use *only* these words throughout the analysis. As Heer (1966) has pointed out, "all intellectual discussions in our time are struggles for language" (p. 474), and he furiously attacks current fashions of playing with and inventing words, rather than trying to define a few with precision. There are no new words in my books—only very old ones that have been used in all societies through all cultures and all time: fear, envy, ambition, trust, friendship, customs, rich, and poor. The ability of my models to define these words and to show the union between words and facts suggests that there is no necessity for inventing new words and gives me confidence that I may be on the right track in the understanding of human behavior. In contrast, the discussion presented so far suggests that many social scientists have just created a body of phrases that, without any justification, implies they have found solutions to problems.

3   . . . and during the Industrial Revolution

Genius may display itself in crime as well as in science, or art, or strategy.

Luke O. Pike, 1873

In the previous chapters I have only examined the consequences of unexpected changes in the distribution of wealth. But one may ask why does the distribution of wealth suddenly change? One of the answers given to this question in my previous book was that accidental fluctuations in population change the individual's share of wealth ("wealth" as defined by the then-existing customs), alter its distribution, and thus shift relative positions. The subsequent gambles on innovations may lead to further accidental changes in the distribution of wealth although, simultaneously, there will be a tendency through innovations made in political institutions to maintain the new order.[1] According to this view, one would expect that periods of great creative activity would be preceded by large, unexpected fluctuations in population, and by a breakdown of customs, by loosening of family ties in particular. These changes lead to greater and more frequent fluctuations in people's position in the distribution of wealth and to an increase in gambling and criminal activities.

The age leading to and including the Industrial Revolution serves as a good example of these views: the brief survey of evidence on these aspects of human behavior will be presented here. This brief discussion, which merely regroups well-known evidence and opinions, serves to set the stage for a detailed account in the next chapter of just one particular response to fluctuation in population size (ultimately an increase) and the breaking down of customs, namely, the innovations made in the English inheritance laws—a type of innovation that, in general, seems to be excluded from discussions on "creativity."

1. Opinions on the Industrial Revolution

The apparently contrasting tendencies of the period preceding the age of the Industrial Revolution and that of the Revolution itself have been noted by various writers. But, with few exceptions, most writers have tended to concentrate on one particular aspect and have failed to put the whole picture together. As a result, Adam Smith in 1776 investigated the causes of increased wealth and recommended laissez-faire instead of the customary restrictions on trade (restrictions inherited unquestioned from the medieval period), but neglected the effects of changes in either the distribution of wealth or population growth on human behavior. Malthus, in his *Essay on Population* in 1798, concentrated on the causes of poverty and did relate them to population growth (although in the 2d edition he significantly revised his views and explicitly noted the role of customs, a point neglected by later interpreters of his works), but he did not look at the increased innovative gambles of the eighteenth century. In 1817 Ricardo, in *Principles of Political Economy and Taxation*, did investigate the subject of wealth distribution and its relation to Smith's system.

Probably the first writer to try to provide a comprehensive picture of that age was Arnold Toynbee, who in 1884 wrote *The Industrial Revolution*, a classic work which gave the period its name. His point of departure is that people's behavior can be explained by the assumption that they fight for a particular *kind* of existence (an assumption which does seem similar to my point of departure) and that men exert themselves in response to external pressures. According to him the difference between those who prefer competition and those who do not is not so much in their understanding of human nature, but in the costs and benefits they attribute to human effort: "Socialists maintain, that [the] advantage [of men exerting themselves] is gained at the expense of an enormous waste of human life and labour, which might be avoided by regulation" (p. 60).

In contrast to Smith, Toynbee also seems to put the customs of the age in the right perspective, rather than viewing them as peculiar burdens. He considers that customs were appropriate for preventing fraud and for controlling the distribution of wealth when the population was smaller and less mobile, but became inappropriate as population and its mobility increased:

. . . the smallness of the world and the community, and the comparative simplicity of the social system made the attempt to regulate the industrial relations of men less absurd than it would appear to us in the present day. [p. 45]

The essence of the Industrial Revolution is the substitution of competition for the mediaeval regulations which had previously controlled the production and distribution of wealth. [p. 58]

But the principle of *laisser-faire* [italics in original] breaks down in certain points not recognised by Adam Smith. It fails, for instance, in assuming that it is the interest of the producer to supply the wants of the consumer in the best possible manner, that it is the interest of the producer to manufacture honest wares. It is quite true that this is his interest, *where the trade is an old-established one and has a reputation to maintain* [italics added], or where the consumer is intelligent enough to discover whether a commodity is genuine or not. But these conditions exist only to a small extent in modern commerce. [p. 56]

Toynbee notes further that when customs are destroyed, the distribution of wealth will change, pauperism and crime will rise, and the appropriate legislation and institutions that will provide some remedies will not be immediately found. At several points in the book he touches on the issue of the randomness of events—in discussing the unintended consequences of the Poor Laws, or in raising the question of whether or not rents are due to chance: "Is [rent] a human institution, or the result of physical causes beyond our control?" (p. 109). However, he neither elaborates this point nor relates rents to the unexpectedly rising population and the breakdown of customs. But one must remember that Toynbee wrote his book at a very young age—he died when he was only thirty years old.

2. Gambling, Social Instability, and the Industrial Revolution

Information on gambling during the eighteenth century is not statistical (perhaps because the population was much smaller then and the information was too well-known to require numerical documentation: the population of England and Wales was less than 6 million in 1700 and only 9 million in 1800). But several surviving documents suggest that a wide variety of games of chance were invented, that there was a large increase in all forms of gambling during this century, and that tickets for the state lotteries were mainly purchased by the relatively poor: John Ashton, who wrote both a history of English lotteries and *The History of Gambling in England* (1898), provides this information.

State lotteries were cancelled in 1826, and Ashton quotes the epitaph written for the occasion:

> In Memory of
> THE STATE OF LOTTERY
> the last of a long line
> whose origin in England commenced
> in the year 1569
> which, after a series of tedious complaints
> Expired
> on the
> 18th day of October 1826 . . .
> . . . it was found that their
> continuance . . .
> encouraged a spirit
> of Speculation and Gambling among the *lower
> classes of the people*;
> [pp. 239–40; italics added]

However, Ashton does not discuss any further the disproportionate participation of the relatively poor in lotteries. Instead he describes in detail several documents that show the increased gambling activity during this age and the incredible variety of gambles that people invented (from a £1,500 bet that a man could live twelve hours under water when sunk in a ship [the ship and the man disappeared], to a bet on the sex of Chevalier d'Eon, the French ambassador to England!)

Ashton makes implicit reference several times to changes in people's position in the distribution of wealth and the widespread gambling:

> The Grub Street Journal of 28 Dec. 1733, gives a practical hint how to utilise Gambling:

> . . . As gaming is becoming fashionable, and the increase of the Poor a general complaint, I propose to have a Poor's Box fix'd up in some convenient Place in every house, which may contain all Money that shall be won at Cards, or any other games. [p. 58]

> Nor was it only in London that this gambling fever existed; it equally polluted the quieter resorts of men. [p. 64]

> It was in the early part of the eighteenth century that betting was made a part of professional gambling, as we read in Smollett's *Adventures of Ferdinand, Count Fathom*. On his return to England, he perceived . . . [that] the spirit of play having overspread the land, like a pestilence, raged to such a degree of madness and desperation, that the unhappy people who were infected, laid aside all thoughts of amusement . . . and risqued their fortunes upon issues equally extravagant, childish and absurd. [p. 155]

It is also useful to mention that Ashton notes both that insurance companies developed rapidly during this period and that there is a similarity between a gamble and insurance. But he neither relates the motivation to insure to the breakdown of family ties and customs, nor sees the difference between the motivation to gamble and to insure, as the following quotation shows: "But, paradoxical as it may appear, there is a class of gambling which is not only considered harmless, but beneficial, and even necessary—I mean Insurance. Theoretically, it is gambling proper. You bet 2s. 6d. to £100 with your Fire insurance; you equally bet on a Marine Insurance for the safe arrival of your ships or merchandise; and it is also gambling when you insure your life. Yet a man would be considered culpable, or at the very least, negligent and indiscreet did he not insure" (p. 275). What Ashton failed to see, though the model in chapter 1 makes it extremely clear, is that by gambling, people have tried either to improve or to restore their status, while by insuring themselves they have tried to protect that which they had already achieved. Both acts are implicitly related to the increased population, its increased mobility, and the resulting fluctuations in people's position in the distribution of wealth: gambling because it may restore some people's status, and insurance because it may either substitute for some broken customs and lost trust, or complement the increased amount of trade.

On Social Instability

To the great despair of historians men fail to change their vocabulary every time they change their customs.

Marc Bloch

While it is possible to be on firm ground when defining and examining games of chance, the ground becomes unstable when one wants to study the crime rate during the seventeenth and eighteenth centuries. The reason for the difficulty is that the word "crime" is not well defined: should riots and acts of sabotage (breaking machines at factories, for example) be included in criminal statistics? If so, why were riots treated differently from other crimes? Why have new words been invented to describe such acts? First, one may tend to include "riots" in criminal statistics. Yet this would not be appropriate, since when the distribution of wealth changes, both laws and customs change as well. Second, the term "breaches of laws and of social norms" is not well defined, and it may not be clear who is committing "the" criminal act: those who import the machinery or those who break it?[2] In order to understand why I shall discuss social instability in general, rather than just criminal statistics, let us first illustrate the ambivalence in the definition of a "criminal act" (even if the legal concept, inherited from a period when circumstances were different, seems unambiguous) by looking at some events from England's history. These events will show that many people were unexpectedly losing their status before and during the eighteenth century (status that was previously kept stable through customs) and will thus help to account for the increased social instability in England during the eighteenth century.

One chapter in Toynbee's (1884) book is called "The Growth of Pauperism," a phenomenon noted by many social scientists who have examined the changes in eighteenth-century England. Toynbee links both this sign of increased inequality and the changes in the Poor Laws to population growth and the resulting breakdown of customs:

> Certain characteristics are permanent in all society, and thus in mediaeval life as elsewhere there was a class of impotent poor, who were neither able to support themselves nor had relatives to support them. This was the only form of pauperism in the early beginnings of mediaeval society, and it was provided for as follows. The community was then broken up into groups—the manor, the guild, the family, the Church with its hosptials, and each group was responsible for the maintenance of all its members; by these means all classes of poor were relieved . . . The "settled poor" in towns were relieved by the guilds, in the country by the lords of the manor and the beneficed clergy. [p. 68]

> The stability of mediaeval society depended on the fixity of all its parts, as that of modern society is founded on their mobility. [p. 70]

[But] it is a great law of social development that the movement from slavery to freedom is also a movement from security to insecurity of maintenance. There is a close connection between the growth of freedom and the growth of pauperism; it is scarcely too much to say that the latter is the price we pay for the former. [p. 69]

By "growth of freedom," Toynbee refers to the breakdown of customs. He attributes the growth of pauperism during the sixteenth century to mistakes made by early legislators who could neither grasp the consequences of increased mobility of labor nor recognize that compulsory provisions for the poor were necessary when customary redistribution was no longer respected.[3]

Similar observations have been made recently by Lis and Soly (1979), who write that as a result of the population growth in the early sixteenth century

The peasants paid a high price for their release from serfdom, for by the sixteenth century the shift from personal to economic dependence was far advanced. [p. 54]

[This] caused the number of landless who could feed themselves only through wage labour to rise in some regions to alarming pro- portions . . . Those with neither sufficient land nor recourse to wage labour . . . could do nothing except migrate . . . The town was . . . a refuge for the poor. [pp. 62–3]

Nowhere [until 1520] was an attempt made to deal adequately with the uprooted poor. [p. 83]

The continuous rise in pauperism during the seventeenth and eighteenth centuries (in spite of a rise in wages and, during the first half of the eighteenth century, a low price of corn) is attributed to several causes that can be implicitly linked to the increased population: the enclosure of com- mon fields and the consolidation of farms, the practice of eviction, and the breakdown of customs, reflected, in part, in the changed attitude of em- ployers toward their employees. In contrast to the old days when employers maintained their men when they were out of work, they now repudiated this responsibility. While one can expect this changed behavior when the population is rising and becoming more mobile, one must raise this ques- tion: who is committng the "criminal" acts, those who no longer fulfill customary obligations, or those who attack the former? The ambivalent answer to this question explains why the evidence described below will refer to all signs of social discontent rather than to the narrowly defined category called "criminal acts."

Population growth, breakdown of customs, growth of pauperism, and increased criminal acts, just like the increased creativity associated with "the Industrial Revolution," did not happen abruptly, but represented a continuous process with its roots in the sixteenth century (or, according to some historians, even before). According to Cockburn (1977a), indicted crime continuously increased during Elizabeth's reign, stayed stable around 1600, and started to climb again after 1623. The indictments of this age (1560–1625) confirm the connection between crime and economic misfortune, and Cockburn summarizes the evidence:

> contemporary writers commonly associated offences against property with those sectors of society thought to have been most affected by the economic and social dislocation of the late sixteenth and early seventeenth centuries—rural labourers driven from the land, unemployed craftsmen, discharged and deserting soldiers. . . . "Weavers' occupation is a dead science nowadays," declared a Colchester weaver . . . "We can get no work," said another, "nor we have no money; and if we should steal we should be hanged, and if we should ask no man would give us; but we will have a remedy one of these days or else we will lose all." Nor did the more general point that crime may be seen as an indicator of social tensions and instability entirely escape [the contemporary writers]. "England had never so much work for a chronicle," [wrote] a commentator . . . in 1616, "never such turnings, tossings . . . in the lives of men and women and the streams of their fortunes. [p. 61]

A similar picture is provided by Bridenbaugh in *Vexed and Troubled Englishmen 1590–1642*, in which he thus describes the great disorder of that age:

> War, enclosures, political grievances, uneven applications of the law, combined with depressions, epidemic diseases, and food shortages, manifested themselves like boils on an otherwise sound body politic . . . With the loosening of family and religious ties, women as well as men sought to sustain themselves by any and every means they could muster; drunkenness and immorality increased along with crime and violence as idleness and mischance forced hapless thousands down into the ranks of the defenseless poor. [p. 356]

The picture from the second half of the seventeenth century and long afterward, according to the detailed account of Pike (and others) in *A History of Crime in England* (1873–76), is hardly different. Pike adds: "it was an instinct with artisans and workpeople of all kinds to attempt the redress of a grievance by violence . . . Imported goods, like machinery, were regarded as a source of injury to the men who had to live by the labour of their hands in England . . . The natural impulse of the sufferers was not to seek a distant remedy by making improvements in their own art and underselling their rivals, but to gain an immediate advantage by destroying their adversaries" (pp. 260–61). Pike also takes note of the fact that highway robberies were an everyday event and that there was an unusually high rate of fraud (in stocks and goods), a phenomenon already noted by

Bridenbaugh (1968) for the first part of the seventeenth century ("cheating and many kinds of fraud seemed on the increase: using false weights or short measures, the quality of leather goods, cloth, was suspect," p. 361). Pike suggests that this and other signs of social instability were a result of the loss of trust:

> The time had come when it was necessary that, if society was to be held together at all, there should be some confidence between man and man, apart from the family tie, and from dependence of retainer upon lord, which were the bonds in earlier ages. [p. 319]

But, in spite of the fact that Pike's book was published between 1873 and 1876, shortly before Toynbee's, Pike does not relate his arguments on the loss of trust and the increased crime rate to the increased population, its increased mobility, or people's fluctuating position in the distribution of wealth. Recently, however, Weisser (1979) suggests that the rise in the crime rate during the eighteenth century can be indirectly linked to both the population growth and its increased mobility:

> The "first cause" of crime . . . was population growth . . . [Also] . . . population movements—numerical and geographical—were crucial elements in the shaping of early modern European crime. If anything this factor would become even more important as the period wore on. For not only did population levels change even more dramatically than ever before but the geographical mobility of the European population would continually increase in numbers and scope. [pp. 107–8]

Reading the documents of the eighteenth century, many other writers have gotten a sense of the greater insecurity and greater disorder of that age. Beattie (1977), for example, wrote:

> Throughout the eighteenth century Englishmen were convinced that crime was increasing. From the anonymous pamphlet *Hanging Not Punishment Enough* (1701) at the beginning of the century to the long and detailed analysis of London crime by Patrick Colguhoun at the end [1796], there was a stream of alarmed comment about the danger of a rising tide of criminality. [p. 299]

> There was an increasing sense not only of the growth of crime but its pervasiveness, a fear that society was in danger of being engulfed. [p. 300]

Pike mentions Fielding's study, *An Enquiry into the causes of the late Increase of Robbers, with some proposals for remedying the growing evil,* which dealt with the first half of the eighteenth century, as a good source of information, although he accepts his conclusion on the increased crime rate with some reservations.

In more recent studies, attempts have been made to look more closely at the figures and see whether or not there was really a large increase in the narrowly defined crime rate. Beattie (1972) first attempts to discover whether the crime rate really increased or whether more crimes were recorded because the state had become better organized. Beattie's conclusion, like that of others, is unequivocal:

> . . . the changes in the administrative system were not so sudden or so massive as to alter fundamentally the sample of crime represented by indictments, to alter it in such a way, that is, as to make indictment totals an invalid guide to changes in criminality. [p. 311]

Radzinowicz (1956) reaches the same conclusion: "the weaknesses in the structure of the machinery for keeping the peace remained at the beginning of the nineteenth century as fundamental as they had been half a century earlier" (pp. 178–79).

Once the possibility that "increased crime rates" could have been due to the better administrative process is eliminated, Beattie (1977) examines the number of crimes against property between 1736 and 1753:

> In the middle of the eighteenth century . . . a crime wave of unexampled proportion struck the metropolis of London and its immediate environs. [p. 155]

> . . . apart from . . . seasonal variations, there was also frequently a sharp difference in annual totals from one year to another. This derives . . . from what appears to be the most pervasive reasons behind property crime—want and necessity produced either by unemployment or falling real incomes . . . there is an apparently close connection between the most striking fluctuations in the number of offenders brought to court and changes in conditions of life and work. [p. 158]

Pike (1873) and Rudé (1972) take note of other manifestations of social instability and look at the rising number of popular protests during the eighteenth century: the destruction of turnpikes, for example, became a frequent occurrence. There were riots around Hereford and Worcester in 1735–36, around Bristol in 1727 and 1753, and in June of 1753 every turnpike near Leeds, Wakefield, and Beeston was pulled down. Enclosure riots were most frequent in 1760 (after the first General Enclosure Act), but they took place in other years, too: in Northampton in 1710, Wiltshire and Norwich in 1758, Northampton and Oxford in 1765, Boston in 1771, Worcester in 1772, Sheffield in 1791, and the Northampton district in 1798. Food riots were frequent, too: of 275 disturbances between 1735–1800 (noted by Rudé 1972), 175 were of this kind, and they occured in "years of shortage and fear of famine (rather than of famine itself) . . . In England there were provincial riots in 1727, sporadically in the 1730s, in 1740, 1756–57, 1766 (the worst year of all), 1772–73, 1783, 1790, and 1795 (Rudé 1972, pp. 287–88). Riots against some laws and minorities were common, to:[4]

> Londoners . . . rioted against Protestant Nonconformists in 1709 and 1715, against Sir Robert Walpole's Excise in 1733, and against the Gin Bill and the Irish in 1735 . . . On a larger scale altogether were the riots on behalf of John Wilkes in the 1760s and 1770s and against Roman Catholics and their supporters in the Summer of 1780. The last of these were the wildest and most destructive and most extensive in the whole of London's history. [p. 291]

Additional signs of social instability have been noted by other writers. In "The Crime of Anonymity," Thompson (1975) shows that the eighteenth century saw the widespread use of anonymous threatening letters and argues that they represent a characteristic form of social protest in which literacy starts to be widespread, customs have been broken, and institutions have disintegrated, while no new ones have yet been found to substitute for them. These threatening letters, addressed to the rich, the authorities, and the employers, were an additional reminder to them to fulfill commitments (protected in the past by customs), if they wanted to avoid violence. In 1730, threatening letters were thrown in houses and workshops in Bristol demanding money on threat of arson, and, after 1790, most letters referring to agrarian conflicts resort to the same sort of threat. Machine breaking and industrial arson were nearly always accompanied by letters, too; according to Thompson (1975):

> A group of such letters comes in the 1780s and 1790s, from the Lancashire cotton industry . . . The most vigorous series comes, between 1799 and 1803, from the West of England shearman and clothing workers. [pp. 274–75]

In 1794 notices appeared in six parishes demanding an advance of wages to the poor: if not, the notices said, the poor will use force. Why didn't they, after all? Thompson (1975) suggests an answer:

> *Given the opportunity*, such insurrectionary voices could be followed by insurrectionary actions . . . But *without the opportunity*, the voices could switch back once again . . . It can be seen in the oscillatory tone of the letter from the commoners of Cheshunt in 1799: on the one hand, the unmeasured violence of language—"Whe like birds of pray will prively lie in wait to spit the bloud" of those preparing enclosure, . . . on the other hand, if instead of enclosure the same gentlemen had effected a fair regulation of common rights, their . . . "will whe come and give our hearts and voices to . . . you for ever." [p. 306; italics in original]

While Thompson does not discuss the meaning of "opportunity," the views presented in the previous chapters make this clear. The term refers either to a leader on whose leadership the poor can gamble, or to a reformer who succeeds in passing a law demanding a redistribution of wealth. In either case the resulting redistribution is not due to a simple, benevolent, act: it is achieved either through violence or under its threat—just like the discussion on progressive taxation has implied. Of course, for later generations the new methods of redistribution become their customs, and the story changes; paraphrasing Pascal and rephrasing Machiavelli's views: customs become a second nature which destroys the first (i.e., the one presented in chapter 1).

3. On Creativity during the Eighteenth Century

Between order and disorder a delicious moment reigns.

Paul Valéry

What association does the term "Industrial Revolution" bring to mind? Does it have a negative connotation and is it associated with suffering and social instability? Or does it have a positive connotation and bring to mind the spectacular creativity of human nature? If popular history books and encyclopedias are viewed as good sources of information for answering these questions, the connotation seems to be the positive one.[5]

There have been an endless number of studies written on the subject of technical progress during the Industrial Revolution (Much less has been written on crime and even less on gambling—no wonder our memories of the past are biased.[6] We remember the triumphs, but who remembers the agonies of an age!)[7] I do not intend to summarize or discuss these studies; they have nothing to say on the question which seems central if one wants to explain the increased creativity of that time: *Why did people gamble more frequently on new ideas?* After all, only by answering this question can one hope to explain why there were more innovations in industries, business, the arts, and legislation. Trying to explain "technical progress" (what's that if not new *ideas* in technology?) in terms like: economic factors ("demand induced," "supply induced," etc.), demographic factors (due to "population growth"—isn't it surprising how inconsistently this factor is used, viewed at times as inducing innovations, at others as causing famines?), psychological factors (being due to "change in taste")—misses the point. After all, events do not belong to categories.[8]

The reason historians and economists writing on the Industrial Revolution present summaries and discussions before presenting their views is to point out the contribution of their study to the existing state of knowledge, while using the customary tools of analysis. But since I do not use customary tools, and since my explanation for the phenomenon has little to do with existing ones, the novelty of the approach comes to the fore in the previous chapters even in the absence of such a summary. A critical and comprehensive one, however, can be found in Mokyr (1984). Yet the views of a few writers should be noted. Boserup (1981, p. 112) argues that population growth led to the Industrial Revolution in Britain. However, she does not show how can there be a link between population growth and innovations. The views presented in this and my previous book show a way to make this link: fluctuations in population, by destroying customs and social institutions, lead to fluctuations in people's position in the distribution of wealth and in the perception of benefits and costs associated with various ideas, and in turn to more frequent bets on new ideas. In contrast to Boserup, Ayres (1944) attributes both technological progressiveness and Europe's unique history to the weaker customs, although he does not link this weakness with the more frequent fluctuations in Europe's population. But the similarity between my views and those of Ayres comes to fore in these lines:

> [Europe] was of all the great civilizations of the time incomparably the youngest, the least rigid, less stifled than any other by age-long accumulations of institutional dust, more susceptible by far than any other to change and innovation . . . Almost certainly it was this composite character which made the civilization of medieval Europe the parent of Industrial Revolution. [p. 137]

Olson's (1982) view is reminiscent of Ayres: according to him, too, the relative weakness of pressure groups sets the stage for innovations (although Olson does not say through what channel such weakness provokes people "to think"). He links the relative weakness of pressure groups to the civil wars of the seventeenth century, which by making Britain into a larger jurisdictional unit, made it more difficult, for a while, for pressure groups to get organized.

   Before briefly summarizing the views on some aspects of creativity during the Industrial Revolution, it may be important to emphasize a methodological point. Explanations of changes concerning human behavior require a causal element which is exogenous, i.e., not under people's control. Accidental fluctuations in population and the resulting random gambling on ideas provide this element throughout all my broader historical examinations—more about their role in the next and final chapters. Let us turn now to a brief summary of various facets of creativity during the Industrial Revolution.

   The eighteenth century seems to be remembered in terms of the almost abrupt progress supposedly represented by the Industrial Revolution rather than as a period of adaptation to Europe's fluctuating and finally rising population and the resulting breakdown of customs. History books give detailed descriptions of Newcomen's engine (1712), Watt's steam engine (1760), the innovations in metallurgy, and the spectacular inventions in the textile trades, from John Kaye's shuttle (1733) to Crompton's spinning mule (1774). They also write about innovations in science, painting, literature, and cultural institutions (under titles like *The Triumph of Culture: 18th Century Perspectives*, 1972), innovations that bring to mind the names of

Hobbes, Locke, Milton, and Newton (from the seventeenth century), and Bishop Berkeley, David Hume, Samuel Johnson, Robert Burns, Daniel Defoe, and Adam Smith (from the eighteenth century). This aspect of the century is so well-known that the only thing I can add to it is that according to more recent research (see Jones 1981, Vilar 1966) these accounts are somewhat biased. The data show that the vigorous inventive activity did *not* in fact start abruptly at the end of the eighteenth century. Rather, its sources were rooted—like those of the increased social instability—in the sixteenth century. Deane (1973), for example, writes:

> As far as British experience is concerned, some writers, of whom Professor Nef is the leader, trace the British industrial revolution back to the late sixteenth and early seventeenth centuries. The period which Nef has picked out as one in which the pace of industrialisation first quickened decisively was the century 1540–1640—beginning with the dissolution of the monasteries and leading to an unprecedented industrial development during the latter half of Elizabeth's reign and the reign of James I. He finds evidence of an enormous expansion in coal-mining and a variety of manufacturing industries, dating from the middle of the sixteenth century until the Civil War, and generating a pace of industrial growth almost as fast as that taking place in the period usually attributed to the first industrial revolution, i.e. between the mid-eighteenth century and the first Reform Act. This upsurge, which was paralleled in a number of other contemporary European countries, was accompanied by a scientific revolution characterised by "changes in the methods of scientific investigation in the direction of controlled experiments and accurate measurement" and by "changes in the goals of industrial enterprise in the direction of quantity production." It had many of the characteristics, that is to say, of what we have chosen to define as an industrial revolution. [p. 165][9]

Other writers, examining the culture and the politics of not only England but Europe, noted the eruption of enthusiasm in general and believed it was not confined to scientific and business creativity alone (see Hughes 1971). Yet, in the eyes of contemporaries, the enthusiasm was also a sign of a stylistic disorder in the spiritual realm and in literature and the arts. The sequence of definitions for the word "enthusiasm" in Samuel Johnson's *Dictionary* illustrates this point (although the Greek origins of the word, meaning "possessed by a God," illustrate it as well—see appendix 1.1):

> The first definition is both the best known and the most hostile: "A vain belief of private revelation; a vain confidence of divine favour or communication." Johnson supports this meaning by quoting Locke's opinion that "*Enthusiasm* is founded neither on reason nor divine revelation, but rises from the conceits of a warmed and overweening Brain." But the second and more neutral definition raises the other sense of inspiration by offering "Heat of imagination; violence of

passion; confidence of opinion." A third definition sounds even more assertive and much more complementary. It calls enthusiasm "Elevation of fancy; exaltation of ideas." [Hughes 1971, pp. 81–82][10]

Here are some other poetic lines typical of that period reflecting an opinion on the new trends:

> A numerous Host of dreaming Saints succeed;
> Of the true old Enthusiastick Breed:
> 'Gainst Form and Order they their Pow'r employ;
> Nothing to Build, and all things to Destroy.
> But far more numerous was the Herd of such,
> Who think too little, and who talk too much.[11]
>
> Dryden; *Absalom and Achitophel*

The same movement against form and order can also be found in British paintings in the twenty-year period between 1760–80. Rosenblum (1971) notes the emotional intensity in the work of James Barry (1741–1809), and at times even in the portraits and paintings of Reynolds (1732–92) in contrast to the "objective" representations of their predecessors. The harmonious landscapes in the tradition of the seventeenth century's Claude Lorrain were also being abandoned: instead Nature rages in the stormy landscapes of Richard Wilson (1714–82), George Barret (1732?–84), and others with lightning and blasted trees, while in Joseph Wright's (1734–97) paintings from Italy, volcanoes erupt and mysterious caverns replace the classical beauties of Italy. The painters of this period create images from literary sources with dramatic themes: King Lear in the storm, the hero of Sophocles' *Philoctetes* isolated in a savage island, and others. The images of "warmed and overweening Brain" are well represented both in the new literary genre—the Gothic novels—and in their translations into paintings: images of hallucinations, sorcery, and magic against a background of mists, moon, and candlelight, and the frequent representation of savage animals that were metaphors for strange, "subhuman" passions abound.[12] These new trends in art and in literature, the abandonment of traditions, the gambles on new forms, new languages that describe the world should not come as a surprise in the light of my views. This is the order—or rather, disorder—of things. When the constraints and beliefs of the traditional order crumble and customs are weakened, people will gamble on new ideas producing economic, political, and social turmoil— electrifying culture at the same time.

In conclusion, this chapter merely summarized additional evidence that is well-known in order to provide the background for the detailed examination of just *one* innovation in England's legal history: the adjustment of its inheritance laws to the fluctuating, and finally rising, population. The examination of the facts will suggest that in this much more narrowly defined domain, my views on human nature are not contradicted either. In addition, the discussion also helps one to understand why it can be very misleading to omit any domain, the legal one in particular, when one writes on the history of human creativity and tries to understand the changed behavior of "modern" man. Social scientists who look either at the whole forest or at a few yellow leaves being carried off by the wind may forget that there are many trees growing and dying there.

4   Why Did Inheritance Laws Change?

Say not you know another entirely, till you have divided an inheritance with him.

J. K. Lavater

Inheritance laws and customs regulate the way wealth is transmitted from one generation to the next. In most societies inheritance laws and customs restricted the power of the individual to leave his property to whomever he wished. In particular, children could not be disinherited (see Goody, Thirsk, and Thompson 1976). Testamentary freedom, and thus the right to disinherit one's children, has been rare. Although it did occur in fourth century B.C. Rome (see Westrup 1936, 2:129), it began to disappear under the rule of Justinian sixth century A.D., who established the "legitim" share of the children. Complete testamentary freedom emerged in England only during the Renaissance.[1]

The existence of restrictive inheritance laws raises a puzzle. Since they restrict the parents' choices, such laws either encourage children to sue their father's estate or provoke conflicts with their siblings. Thus, at first sight, it appears that these laws may be costly for society. If so, the question that immediately arises is why these laws survived for such long periods of time? Why weren't they altered once their harm was perceived?

This chapter shows that these restrictive laws have benefits as well as costs, since they diminish inequality among heirs and thus enhance the stability of the society. Their advantages are evident when people live in relatively isolated communities, where central authority is weak. When population increases and central authority becomes stronger, the stability of the society is enhanced if inheritance laws are *changed* and greater freedom of choice is given to parents. The evolution of inheritance laws and customs in England and France during and after the Middle Ages documents these conclusions.

1. On Egalitarian and Nonegalitarian Inheritance during the Middle Ages

The explanation frequently given for the existence of restrictive inheritance customs is that they are "fair." Yver (1966) and Le Roy Ladurie (1976), for

This chapter was written by Gabrielle A. Brenner.

example, rely on quotations from the customary laws of diverse French regions:

> For the custom is . . . that no person . . . may make the condition of his heirs presumptive worse or better of one than the other. [Customary of Maine]
>
> Parents [should not] make one of their heirs better than another. [Poitou Law][2]
>
> The father or mother cannot advantage one of their children in their succession more than the others. [Paris Law][3]

The views presented in chapter 1 support these explanations for restrictive inheritance customs and thus provide a theoretical background for their existence. Less inequality among heirs diminishes the probability of conflicts occurring among them, thus increasing the stability of both the family and the society.

One may immediately ask why inheritance laws today are so different, at least in the Anglo-Saxon countries. Why do they no longer impose restrictions on parents? The answers suggested in this chapter attribute the change to several interrelated factors: increased life expectancy, increased population, and the substitution of customs by laws and police backed by strong central authority.

On Egalitarian Inheritance Laws

In the Middle Ages the probability of a father dying before all his offspring were grown up was relatively high. This information can be derived from table 4.1 which shows the life expectancy at adult age among members of English ducal families. In the fourteenth century, the twenty-year-old expecting to die of a nonviolent death could look forward to living only until the age of 51. If violent death was taken into account, the expected age fell to 42. These numbers refer to groups whose average nutrition and health were probably better than that of the average population. If fathers died at such relatively young ages and if no restrictive, egalitarian, inheritance customs existed, the elder brothers and sisters would no doubt have precluded younger siblings from receiving any significant part of their father's estate. The reason for the elder sibling's attitude is that in a sparsely populated society with a weak central government (that is, in most medieval societies), family and kin provided insurance against aggression and support for old age (see Bloch 1940, Brenner 1983b). But, although one's younger brothers can thus eventually provide such insurance, one's children may be expected to do so even better. Thus to exclude one's siblings to the ultimate benefit of one's children is a strategy that may increase the probability of either maintaining or ameliorating one's position in the distribution of wealth. But such behavior, if unchecked by custom would impose a responsibility on the rest of society which either would be left to provide for the penniless orphans or would suffer from the consequences of orphans gambling on a life of crime. In the absence of restrictive inheritance laws or customs, such societies would thus be left either with an increased crime rate, or with claims that would strain its charitable institutions (i.e., with effects called "externalities").

The absence of restrictive inheritance customs would have had further costly consequences due to an additional feature related to the relatively young age at death. Family ties and expectations of insurance are expected to be weaker between stepbrothers or stepchild and stepparent than between either full brothers or children and parents. The restrictive inheritance customs protected the stepchild from being totally disinherited by his stepbrothers, and society from having either to support him or suffer his depradations if he turned to crime.

While little numerical evidence has survived to support these views, there does exist a wealth of nonnumerical evidence that does not contradict them. This evidence shows that the family provided more insurance in medieval societies than today, that parents did die at relatively young ages, and that the twin phenomena of "cruel" stepparents and of orphans might well have been common. Indirectly, this evidence supports the view that restrictive inheritance customs represented adaptations made by medieval society to increase its stability.

The greater role of the family in providing insurance during the Middle Ages has been emphasized by many social scientists (see, for example, Nelson 1949 and Bloch 1940). Generally, it has been argued that the concept of property was different then from what it is today, "wealth" being viewed at that time as a property of the extended family and not of any of its individual members, and the father seen as the chief caretaker and manager of this common property (see Bloch 1940; Westrup 1936, p. 29; Pollock and Maitland 1898, p. 248; Brenner 1983b, chapters 2 and 3). This insurance pattern has been attributed to the drastically diminished population after the heyday of the Roman Empire[4] (see Russel 1972 and Bloch 1940), and is a general characteristic of sparsely populated, preindustrial societies. Goody (1976), for example, remarks that

> . . . Security in old age . . . is a prime consideration when there are no alternative means of saving up for the period when an individual can no longer support himself . . . ; in these situations an important part of one's savings, one's capital, consists of one's kin . . . [p. 87]

Becker (1981, pp. 238–39) makes similar points: he mentions that considering kin as "insurance agents" is a good strategy when life and marriage spans are expected to be short, and Bloch (1940) describes some types of insurance provided by kin in feudal times:

The man who was brought before a court found in his kinsmen his natural helpers. [p. 124] In the vendetta . . . the ties of kinship showed themselves at their strongest . . . The whole kindred . . . took up arms to punish the murder of one of its members or merely a wrong that he had suffered. [pp. 125–26]

while Flandrin (1979) says that

In the tenth and eleventh centuries, in a society in which the royal power had become almost non-existent, the *lignage* had had, as an essential function, the protection of its members. [p. 15]

The information on the shorter life expectancy during this age can be inferred both from table 4.1 and from additional sources. Russel (1972) found that nearly half of the sons in medieval times inherited from their fathers while minors, another indicator of the young age at which their fathers died, and between 1377 and 1485 three English kings (Richard II, Henry VI, and Edward V) out of seven acceded to the throne while minors after their fathers died a nonviolent death. Also, as table 4.2 shows, remarriages following the death of one of the partners were frequent: among the British peerage, for instance, the proportion of second marriages was 36% between 1550 and 1574. Since before 1550 the death rates were higher (due to epidemics like the Black Death), this proportion of second marriages must have been even greater then. Moreover, the peerage is probably a biased sample of the population,[5] and the rate of remarriage among the general population could have been different.

If all these marriages and remarriages resulted in live offspring, the children of a significant number of families ended up having a stepparent, stepbrother (or stepsister), or both. Some nonnumerical evidence is suggestive: Iona and Peter Opie (1974), in their introduction to fairy tales, explain the prevalence of the stepmothers in them by "the shortness of life in past times, by the consequent shortness of marriages, and by the practice of the surviving partner marrying again without unnecessary delay" (p. 19). Moreover, the fairy tales the Opies examine frequently refer not just to a stepmother or stepsisters but to a *wicked* stepmother (or wicked stepsisters) trying either to kill the stepchild or to steal his inheritance. Cinderella and Snow White are only the most famous examples of this kind of tale, but there are many others whose stories are similar.[6] Darnton (1984) found that in most European folktales younger sons and stepdaughters are very common underdogs (p. 55), while Goody (1976) noted that

The prime example of the step-situation, that is, the Cinderella story, is largely a European tale. . . . Cinderella could not have been invented in a society without step-mothers . . . [pp. 54–55]

and Smith (1953), commenting on the enduring elements in fairy tales, says that

People are interested in and sympathise with those things they themselves have experienced. When the story teller told about the stepmother and her ill-treated step-daughter, usually the idea must have been present in some degree, or it would have dropped out of the story. [p. 19]

According to Darnton (1984) the reason why stepmothers rather than stepfathers are so prevalent in fairy tales may be that widowers remarried twice as frequently as widows (p. 27).

On fairy tales in general, Mueller (1983) has pointed out that those retold by the Grimm brothers constituted the lore of law in Germany's remote regions. Tales of burned witches, abandoned children, or murderous, wicked stepmothers are not exactly bedtime stories (the only reason we fail to see their violence and cruelty is that we are already so accustomed to them). They should be seen as an education in law for the younger generation, an education that in the sixteenth, seventeenth, and eighteenth centuries was still mainly transmitted orally, since many people were still either illiterate or barely literate. To transmit the law in this way therefore made sense. As Mueller puts it, these aren't really "fairy" tales but "fair" tales, for in all these stories, the "fairy" is usually the dispenser of fairness (as it was understood then).

This observation on the relations between stepparents and stepchildren is not new, as these lines from Juvenal (1st century A.D.), quoted in Goody (1976) suggest:

Wives loathe a concubine's offspring. Let no man cavil
Or call such hatred abnormal: to murder your step-son
Is an old-established tradition, perfectly right and proper.
But wards with rich portions should have a well-developed
Sense of self-preservation. Trust none of the dishes at dinner:
Those pies are steaming-black with the poison Mummy put
    there.
Whatever she offers you, make sure another person
Tries it out first: let your shrinking tutor sample
Each cup you're poured.

[6:626–34]

Evidence on the mistreatment of children after remarriage is also provided by Cooper (1976) who remarks on inheritance practices among England's nobility: "The extreme cases of depriving eldest sons for the benefit of younger ones usually came about as a result of second marriages" (p. 215).[7] Numerical evidence on the cruel behavior of stepparents comes from a later age: Cockburn (1977), in his analysis of the crime data of sixteenth-century Essex, notes that

A surprisingly high number of family killings—ten of the nineteen recorded—involved the murder of sons, daughters or step-children

. . . A clue to the possible motivation . . . lies in the scramble for property and inheritance which underlay the brutal killing of an Essex boy in 1595 by his step-father and brother-in-law. [p. 57]

In conclusion, the restrictive, egalitarian inheritance customs during the Middle Ages can be attributed to the decreased population and its relative lack of mobility, to the resulting stronger blood-family ties, and to the unflattering view of human nature illustrated by the model in Brenner (1983b), a nature now revealed in the (mis)treatment of both stepchildren by stepparents and younger children by their older brothers and sisters.

On Nonegalitarian Inheritance Laws

The laws governing noble fief inheritance were often different from those for the rest of the population (see Bloch 1940, pp. 203 ff.; Pollock and Maitland 1898, 2:262; Lefèbvre 1918, 2:13). The noble fief was not partible, but it was frequently devoluted upon the eldest son. In England, male unigeniture did apply to all land held in "military fiefs" (defined as those fiefs liable to pay the war tax or scutage; see Pollock and Maitland 1898, 2:262). In France, male primogeniture had been the rule for the transmission of all "military" or noble fiefs since at least the ninth century (Lefèbvre 1918, 2:200–201).[8]

This difference between inheritance among fiefs and other inheritance practices seems, at first sight, to contradict our explanation. In these times the possession of landed fiefs brought more honor and power than any other form of property, since almost all social standing and wealth were derived from the land. Thus the unigeniture of noble fiefs led to *greater* inequality among noble sons. But, according to our view, this practice was bound to compel society to deal with the upheavals the penniless younger son of fief-holders might cause (as indeed it had to: according to the testimony of a Continental visitor to England in 1584, "the eldest sons of the nobility inherited all, while the younger took up some office [became entrepreneurs?] or pursued highway robbery" (as quoted by Thirsk 1976, p. 185).[9] Yet an explanation may be offered for the difference in practice between the two classes.

Recall that during the medieval period the fief-holders provided the military insurance. If the fief had been split among the heirs, military security would have been diminished. Thus, when chosing the inheritance pattern for nobles, two types of risks had to be taken into account: the greater risks of being attacked by outsiders if the fief was to be split among heirs, and the greater risks of internal instability if the inheritance was nonegalitarian and the fief was transferred to just one son. Apparently, as the evidence below suggests, the latter risk was generally considered to have less severe consequences, and led to the nonegalitarian inheritance pattern among fiefs for sons. It should be added that when only daughters, who presumably were married to other members of the warrior caste, were left, leaving the fief to just one of them might have led to disruptive feuding among the husbands of the sisters. This may have resulted in the widely followed egalitarian practice of splitting the holding among daughters if no son was left to inherit.

The evidence is in fact consistent with these views. Historians have singled out the role of protector of the feudal lord: Bloch (1940) explains the origin of the tie between lord and vassal as the need for protection when *neither the State nor the family any longer provided adequate protection* (p. 148) after the collapse of the Roman Empire. North and Thomas (1973) explain the specific arrangement of the manorial economy to be due in part to the fact that ". . . the fortified castle and armored knight on horseback, having specialized skills in warfare, provided local security which could never be equalled by . . . peasants" (p. 19). Primogeniture has been linked to the provision of this insurance: Lefèbvre (2:200–201) attributes it to the fact that the elder son was the most likely warrior to defend the fief in troubled times. Blackstone (1766, 2:215) attributes it to both the weakening that resulted from splitting up the estates and the incentives it provided for younger sons to be "serviceable to themselves and the public, by engaging in mercantile, in military, in civil or in ecclesiastical employments" (p. 215), in other words to become entrepreneurs. This is the same argument given by Guy Coquille, a sixteenth-century French jurist (quoted in Lefèbvre 1912, 1:144) who justified the advantages given the eldest son by saying that if the estate was split, it would greatly reduce its wealth, while with primogeniture the youngest sons had an incentive to seek their fortunes elsewhere. Holdsworth (1903, 3:173–74) reports that in the thirteenth century primogeniture was "said to be needed in order to maintain a race of wealthy landowners who can see to it that the land is cultivated and the cultivators protected." Russel (1972, p. 49) also attributes primogeniture to attempts to avoid conflicts among sons and concludes that the fight among minor heirs for the (partible) crown was one of the main reasons for the fall of the French Merovingian and Carolingian dynasties:

> The Merovingian dynasty was plagued by many minorities as well as by equal division among sons, since, on the average, two sons survived. The later Carolingians also went down under the weight of equal division among heirs and minors. The early Carolingians . . . were saved . . . by one of the most remarkable instances of operation of chance in history—for five generations they inherited at about age 25 and lived long enough to be succeeded by only *one* adult son (either because he was the sole surviving son or because only one wished to rule). [pp. 48–49; italics added]

As a result of this chance happening, according to Russel, the next dynasty of French kings, the Capetians, saw the benefits of one heir and adopted primogeniture.[10]

Other writers have noted the harm inflicted by partible inheritance customs. As far back as 1584, David Powell argued that partible inheritance customs had destroyed the Welsh nobility by, among other causes, "exacerbating strife among brothers" (Thirsk 1976, p. 188). Leyser (1968) attributes the feuds and rebellions of the early medieval German aristocracy to *the system of partible inheritances from which neither kings nor dukes nor lesser power could depart* (p. 38). Holdsworth (1903) quotes a 1276 charter of Edward I:

> It has often happened by the ancient Kentish custom of partition by gavelkind that lands and tenements, which in certain hands when undivided are quite sufficient for the service of the state and the maintenance of many, are afterwards divided and broken up amongst coheirs into so many parts and particles, that no one portion suffices even for its owner's maintenance. [3:174 n. 1]

Sabean (1976) remarks that "Within a military caste disputes among sons and between fathers and sons were apt to be too disruptive. Those sons without expectation of an inheritance . . . were forced to leave home" (p. 100). Duby (1973) wrote that not only younger but also elder sons could provoke quarrels on inheritance and were expected to go and look for heiresses while their father was still alive.[11] Moreover, he remarks that a lot of eleventh-century younger sons left for the Crusades mainly in search of wealth (1983, p. 105) and that one strategy used by feudal noble families to avoid too many conflicts among sons and family members was to let only the elder marry: the younger sons either became soldiers in search of an heiress to marry or monks (pp. 43, 146, 268).

We may also note that the unigeniture inheritance system of the warrior class of fourteenth-century Japan has been linked by Nakamura (1981) to the need to provide security when central authorities were weak:

> The essence of the *ie* system which originated in the Ashikaga period (1333–1558), was that *ie* (household) assets would not be divided but would be inherited by one heir. This institution was the only way for a feudal lord to insure that his designated heir would remain economically and militarily strong enough to survive in that period of civil strife, including the eighty-year period known as the Sengoku Jidai, the period of the warring states (1477–1558). The division of a domain among many sons would mean that each would be at the mercy of neighboring predatory lords. [p. 273]

While the aforementioned evidence showed the benefits of primogeniture (or unigeniture) among fiefs, in providing military insurance, its harm was apparent at times, too, due to the internal strife caused by the nonegalitarian practice: Bloch (1940, pp. 134–135) gives examples of noble brothers killing and dispossessing each other and mentions both the infighting for the Crown and the frequent murders in the family of the Angevin kings of

England, which he calls the "Atrides" of the period. Duby (1983) gives examples of upheavals caused by younger sons in the noble families of Catalonia and Anjou in the eleventh century where *youths [were] expelled and forced to go far away . . . Brothers slew one another . . .* (p. 63). Cooper (1976) mentions that very long litigations may have resulted from the practice of primogeniture and that in many cases the Crown was forced to arbitrate through fear of local disturbances caused by parties who tried to settle their quarrels over inheritance by force. In addition to the aforementioned evidence from Thirsk (1976) on younger sons becoming highway robbers, he also remarks that "The literature on primogeniture in England . . . [began to accumulate when people] . . . experienced the practical problems of younger sons of gentlemen almost begging in the street . . ." (pp. 185–86).

In conclusion, the differences in inheritance customs between nobles and ordinary people may be attributed to the military role played by the former in feudal societies.

2. Population Growth and the Change in Inheritance Customs

Inheritance customs have slowly changed since the fifteenth century. Two questions are examined here: Why did they change, and is the direction these changes have taken consistent with the prediction that parents would ultimately receive greater power?

Europe's population and its people's life expectancy have grown continuously since the fifteenth century. Table 4.1 showed the rise in life expectancy during this period: from 1480 to 1779 the life expectancy of an English ducal family member at age 20 changed from 26 to 40 years.[12] Hollingsworth (1969) and Hatcher (1977) present evidence on the new trends and attribute them, among other factors, to the drastic decrease in the frequency of plagues, and Wrigley and Shoffield (1981) estimate that England's population grew from 2.7 million in 1541 to 5 million in 1701. According to McEvedy and Jones (1978) trends were similar in all of Western Europe.

Just as the rise of feudal societies has been associated by some historians with Europe's depopulation (Bloch 1940), their fall has been attributed by others (North and Thomas 1973; Brenner 1983b) to increased population. The arguments that link population growth with changes in social structure may thus be briefly summarized: when people live in relatively isolated, small communities, with the expectation of a stable population, religious and other customs may be sufficient to carry out intertemporal exchanges and protect insurance and oral contracts. Written contracts and protection by laws and police are less needed—trust and kin substitute for them. But when population increases, the frequency of interaction between any two individuals diminishes. At that time written contracts, police, and laws backed by strong central authority will offer protection that was previously provided by trust and custom alone. Expectations of a diminished frequency of interaction also imply that kin ties will be weakened, since it is now easier for kin to refuse to fulfill commitments previously based on customs. Moreover, the simultaneous rise in the power of the state decreases the necessity to rely on one's kin to insure one's life and property.[13] The old inheritance customs are put under strain by these changes since they depend, like most medieval institutions, on the relative lack of mobility of the population, and since children can no longer be expected to provide the extensive insurance that they provided before.

Thus, unsurprisingly, the words "family," "friends," and "children" started to take on a new meaning at the end of the Middle Ages. Neither children nor friends were expected to provide insurance to the same extent as before, nor was the (extended) family expected to provide them. Moreover, with increased life expectancy, the probability of a father surviving until his children were adults increased, and that of a child acquiring a potentially hostile stepparent diminished, lessening the probability of a younger son being dispossessed by his elder siblings because he was too young to defend himself. Thus this benefit of restrictive inheritance laws to shield stepchildren from their elder brothers or stepparent lost most of its appeal.

The ancient custom of restricting the freedom of fathers to leave an inheritance at will started to work *against* the fathers who could no longer compel support or obedience from their children by the threat of disinheriting them (is this one of the lessons Shakespeare wanted to teach his audiences with King Lear's story?). Thus society was now threatened with the risks of having to provide for the father and of having sons abandon established ways. As a consequence, the incentives to gamble on new patterns of inheritance which gave more power to fathers increased. Let us examine how this transition, or betting process, occurred.

The decline in the role of the family as insurer in the sixteenth century has been well documented. Stone (1979) relates the process both to the rise of the central state and to the increase in population:

> In sixteenth-century England, rapid demographic growth, . . . urban immigration . . . meant that support from the extended network of the kin and from neighbours . . . became inadequate for large numbers of orphans, widows, cripples, sick, and aged. . . . Welfare . . . had . . . to be . . . taken over by public bodies. [pp. 106–7]

Flandrin (1979) and Nelson (1969) make similar points. Flandrin writes that the "ties of the lineal solidarity became weaker as regards both vengeance and rights to landed property" (p. 25), and Nelson notes the decline in the obligations of friendship. But Flandrin (1979) also notes the simultaneous rise in the power of fathers over their children:

It seems that the authority of parents and their power of coercion on their children increased from the sixteenth century onwards . . . Fathers . . . preserved or recovered . . . most of the rights granted them by the ancient Roman Laws. [p. 130].

Writers of the sixteenth century did recognize some of the changes that were taking place in the perception of the roles of children. The extremely successful 1530 treatise *On Civility in Children* (De civitate morum puerilium) by Erasmus of Rotterdam is evidence of that. Not only was this book reprinted thirty times during Erasmus's lifetime, but it was also imitated many times, sequels to it were written, and it was translated into English, German, Czech, and French (in the latter four translations were made, in 1537, 1559, 1569, and 1613). Although the book is written for the instruction of boys in a society in transition, it reflects the traditional approach to children. Norbert Elias (1978) noted:

Erasmus' treatise comes at a time of social regrouping. It is the expression of the . . . transitional period after the loosening of the medieval social hierarchy and before the stabilizing of the modern one. [p. 73]

To us, Erasmus's treatise may seem strange: he discusses neither childhood nor how children ought to behave. Rather, he shows how children must be educated to enter the adult world. The distance between the behavioral and emotional standards of adults and those of children was relatively small; children, judging from the treatise, were treated as adults from a relatively early age.

But the sixteenth century also shows the first signs of the development of the idea of "childhood" as a particular and recognizable stage of life; children began to be dressed differently and to be more pampered: childhood was now being viewed as a longer period associated with dependence and "immaturity" (Ariès 1973, pp. 62–75). If our views are correct, this should not be surprising: when children could no longer be expected to fulfill customary commitments because of the family's smaller control on them, and when the state was not yet strong enough to substitute for all the insurance previously supplied by children, building stronger sentimental ties could have provided one means of insurance during this "transitory" period (that still continues today?).

As to the change in inheritance laws themselves: from the fifteenth century on, an attempt was made in England to sidestep the strict common-law rule governing the disposal of one's landed property by testament, by adopting a device, the "use," which in effect allowed fathers to dispose of landed property by will.[14] This device, however, led to widespread uncertainty about property rights, since buyers did not know whether the property offerred to them for sale belonged entirely to the seller. In order to correct for the abuses and frauds caused by this uncertainty, and yet provide the father with greater freedom of choice, the Will Act of 1540 allowed the free disposal by Will of all land held in socage tenure and two-thirds of that held in knight service on which no restriction had been imposed by the original grantor.[15] (This restriction of two-thirds seems to have been imposed for political reasons and was repealed under Charles II, in the seventeenth century).[16]

In the late fifteenth century, lawyers found a way to break entails, a device that restricted the freedom of the head of the family to dispose of his property (Stone 1979, p. 112).[17] A 1540 statute weakened the entails even more (Cooper 1976, p. 203), giving the head of the family additional freedom to dispose of his estate. Stone notes:

> This [change] greatly strengthened the ability of the current head of the family to dispose of the property as he chose . . . [It] meant an increase in his capacity to punish or reward his children . . . He could threaten them with total exclusion from the inheritance. [pp. 112–13]

Finally, in 1597, a statute outlawed another kind of restriction in the freedom of will of the father, the perpetuity. Francis Bacon, the lord high chancellor, gave the following argument for its abolition:

> Though I reverence the laws of my country, yet I observe one defect in them; and that is, there is no footstep there of the reverend *potestas patrias* . . . This only yet remain: If the father has any patrimony and the son be disobedient, he may disinherit him . . . But this device of perpetuities has taken this power from the father . . . [quoted in Holdsworth 1905, 7:201]

and Coke and Popham (Cooper 1976, p. 206) gave a similar interpretation of this change.

When we consider the strategies of other countries for changing the inheritance customs, we are navigating in a more uncharted territory: France, for instance, was united only after the Revolution, and each region, even each town, had its own customs. Thus, as explained before, the incentives to change the inheritance customs were lessened. Nevertheless, there seems to have occurred there in the later Middle Ages "a widespread tendency for the head of the family to acquire greater powers of testamentary disposal of his property, ignoring customary rules" (Cooper 1976, pp. 198, 265–66). In Burgundy there was a revision of the rules in 1570 which gave more latitude to the father in disposing of his estate (Lefèbvre 1918, 2:75), and, as in England, the royal ordinances of Orléans in 1560 and Moulins in 1566 restricted the creation of both entails and perpetuities (Lefèbvre 1912, 1:170). But, because there were many customs, it seems to have been an easy statute to breach, and it did not give the father as wide powers as the English statute did (Cooper 1976, pp. 263–65). Another device that gave fathers a greater power of testamentary disposal was the slow acceptance in the sixteenth century of the right of the father to disinherit an unworthy child completely when he had just cause, *just* being nevertheless restricted to a few well-defined reasons such as marriage against the wishes of the father (see Lefèbvre 1912, 1:174–79).

Thus in both France and England, society gambled on new inheritance customs which did increase the father's rights of disposal of his property as a means of ensuring a greater control over the children when customs could no longer force them to provide insurance. The new inheritance laws substituted for the loosened customs.

Conclusions

This chapter has shown that restrictive inheritance customs benefit societies with stable populations, but do not benefit those with larger and more mobile ones. The arguments have shed light on English laws during the Middle Ages and on changes in them toward the end of that period. Moreover, there is evidence to suggest that sixteenth century France moved in the same direction, although, in contrast to inheritance laws in England, those in France did not give complete freedom of bequest.

One may thus ask why French law did not give complete freedom of will to the father. Some speculative answers can be given: Before the Revolution, each region in France, while being part of the same kingdom, had its own laws and its own judiciary system, in contrast to England where the judiciary system was already unified by the eleventh century. Thus it may have been more difficult to change the laws in France. Second, France had been bled white during the fourteenth century by the Hundred Years War (1328–1461),[18] and later its population growth was slower than that of England.[19] But, in each historical process (as the views in chapter 1 suggest), chance may have played a role. Thus we do not pretend to explain all the details of the changes that took place—only part of them.[20]

Table 4.1 Expectations of Life at Adult Ages (All Deaths)

Cohort born	1330–1479	1480–1679	1680–1729	1730–1779	1780–1829	1830–1879	1880 1934 (etc.)
Males							
age 20	21.7	26.3	30.0	39.9	42.7	39.8	39.8
age 40	13.1	18.3	22.4	25.7	27.0	27.2	29.4
age 60	10.0	9.2	13.2	12.7	13.3	13.4	14.7
Females							
age 20	31.1	29.1	35.4	44.2	44.8	46.2	54.3
age 40	19.2	18.3	24.9	29.9	32.8	31.5	37.4
age 60	8.2	10.3	12.3	16.1	17.6	16.6	21.2

Expectations of Life at Adult Ages (Nonviolent Deaths Only).

Cohort born	1330–1479	1480–1679	1680–1729	1730–1779	1730–1829	1830–1879	1880 1934 (etc.)
Males							
age 20	31.5	30.5	32.8	41.3	44.1	41.9	49.4
age 40	18.7	19.9	23.0	26.0	27.4	28.0	31.8
age 60	12.3	9.2	13.3	12.7	13.3	13.4	18.1
Females							
age 20	31.1	29.6	35.5	44.2	44.8	46.4	54.9
age 40	19.3	18.5	25.1	29.9	32.8	31.8	37.4
age 60	8.4	10.3	12.6	16.1	17.6	16.6	21.2

Source: Hollingsworth (1957).

Table 4.2 Proportion of Second, Third, Fourth, and Fifth Marriages in Relation to the First: British Peerage

Cohort born	First marriage	Second marriage	Third marriage	Fourth marriage	Fifth marriage
1550–74	1.00	0.36	0.09	0.012	0.00
1575–99	1.00	0.23	0.04	0.005	0.001
1600–1624	1.00	0.22	0.03	0.005	0.001
1624–49	1.00	0.22	0.04	0.005	0.00
1650–74	1.00	0.20	0.03	0.005	0.00
1675–99	1.00	0.16	0.02	0.003	0.00
1700–1724	1.00	0.14	0.02	0.001	0.00
1725–49	1.00	0.17	0.01	0.001	0.00
1750–74	1.00	0.16	0.01	0.0008	0.00
1775–99	1.00	0.12	0.01	0.0007	0.00
1800–1824	1.00	0.12	0.01	0.0005	0.00

Source: Derived from Hollingsworth (1964).

5   On Politics and Inflation

Politics: Who Gets What, When, How
                              Title of a book by Harold D. Lasswell

Traditional economic analysis, sometimes with good reason, divides the problem of inflation into two distinct subquestions. The first concentrates on the relation between monetary growth and inflation, while the second inquires into the political economy of monetary growth, that is, the question of why central banks decide to follow a particular policy.[1]

The recent "rational expectations" models analyze the first question. It is the purpose of this chapter to analyze the second by relying on the model presented in chapter 1, and to attempt to answer these questions: Why do central banks behave the way they do? What conditions are more likely to lead to large and sustained monetary accelerations?[2]

There is a reason for trying to provide a uniform approach to the second question. The now-popular "rational expectations" models are based on the premise that unless unexpected changes occur in monetary policy (or in government policies in general) things are bound to change for the "better." But, if such is the case, one must ask, why would somebody capriciously increase the money supply?

This chapter shows that decision makers within governments and central banks are more likely to be driven to change policies not when everything goes smoothly, according to expectations, but when the economy's wealth diminishes (or when conditions, though getting easier, may still fall behind rising expectations), or when the distribution of wealth changes significantly. In these cases the incentives in the economy are changed and people become more likely to put various forms of political pressure on the government in order to redistribute the national pie in one way rather than in another. It is in these circumstances that inflationary policies are more likely to be adopted. A discussion of how the views of Keynes and the Keynesians fit into this view will, I believe, yield a surprise.

In the first section the model presented in chapter 1 is adapted to discuss the relationship between changes in the distribution of wealth and political pressures on governments and central banks to pursue inflationary policies. Next it is shown that the approach presented here is not as novel as it may seem at first sight and that some of Keynes's views on the labor market and on uncertainty, both in the *General Theory* and elsewhere, seem to be similar to those presented here. The third section shows that a new interpretation can be given to both some postulated "macroeconomic" relationships and to some empirical findings, such as Phillips curves and business cycles of various lengths and magnitudes.

1. What Happens When Realizations Run Behind Expectations?

As was shown in chapter 1, when an individual unexpectedly either does not reach his expected position in the distribution of wealth, or loses it, it is more likely that he will start gambling on new ideas, that is, become an entrepreneur and make greater efforts. It has also been shown that when the distribution of wealth unexpectedly changes, it is more likely that people will gamble on new strategies that may have the consequence of redistributing wealth in their favor.

The second strategy may take many forms; the group whose wealth has relatively diminished may lobby for tariffs, regulations, and direct subsidies in order to redistribute wealth in its favor. Increasing government expenditures in its favor (by an expansionary monetary policy in particular), may be just one of the strategies to which governments may resort.

There is a difference between the first strategy and the second one: while the first strategy may increase society's wealth, the second does not always do so; it may only redistribute wealth and, simultaneously, have the following additional effect. Suppose that a group has been successful in reallocating wealth through tariffs, regulations, or taxes. Such success may be perceived by other groups in the society as implying that, were the circumstances to demand it, it would be more profitable to be organized in larger groups than to gamble on individual effort. The changed incentives have the effect of diminishing society's wealth, since efforts will be made to get organized in larger groups rather than in other directions.

How long it takes from the moment such policies of redistribution are implemented until the moment that they are perceived as significantly diminishing the total pie, one cannot say. Consider the model presented in chapter 1: suppose that one group succeeded in its strategy to redistribute wealth in its favor, and let that sum be $100 million. If there are 50 million people in the labor force who do not belong to this group, the redistribution will cost each of them, on average, $2. Such amounts do not alter one's behavior toward risks in my model. But, if such processes of redistribution are expected to be frequent, at one point the sums may add up to a relatively large sum, in which case people's attitudes toward risks may change. When this happens, it will be more likely that policies will be revised, groups will try to redistribute wealth in their favor, and inflationary policies will more likely be adopted.

Governments (or federal banks) are also more likely to resort to inflationary policies when, due to an external shock (for example, an OPEC-type price increase), the wealth of some groups becomes significantly less than expected. Since these governments will have made commitments before their revenues unexpectedly decreased, changes must be made in their methods of financing that will necessarily disappoint the expectations of some group in the economy. Thus incentives have been changed and the probability that pressures will be put on these governments to pursue inflationary policies increases. In this, as well as in the previous case, inflationary policies (through monetary expansions) may succeed in redistributing wealth to the benefit of a particular group and make it more difficult to realize what actually is happening: the noise caused by the expansionary policy itself is mixed with the one caused by the struggle over the redistribution of wealth.

The Redistribution of Wealth in Labor Markets

There are implicit and explicit methods that can be used to redistribute wealth. The explicit ones are, for instance, lobbying for import quotas or imposing differential taxes on sectors, subsidizing one sector rather than another, or raising various barriers to entry. The implicit strategies can be more subtle, and they are the ones that may lead to inflationary policies.

Consider Keynes's (1923) pre-Keynesian, then traditional, description of inflationary processes after World War I in the United Kingdom, U.S., France, Italy, Germany, Canada, Japan, Sweden. He starts his analysis by asking: "How have the price changes of the past nine years [1915–23] affected the productivity of the community as a whole, and how have they affected the conflicting interests and mutual relations of its component classes?" (pp. 3–4).

Keynes divides his arguments into two parts: first, he looks at the change in the distribution of wealth between wage earners and other groups during periods of inflation, and second, he looks at inflation as a method of diminishing the value of the outstanding national debt. Let us examine his views on the first issue here, and on the second in the next section. Keynes summarizes the question of the redistribution of wealth in the labor market as follows:

"It has been a commonplace of economic textbooks that wages tend to lag behind prices, with the result that the real earnings of the wage earner are diminished during a period of rising prices. This has often been true in the past, and may be true even now of certain classes of labor which are ill-placed or ill-organized for improving their position. But in Great Britain, at any rate, and in the United States also, some important sectors of labor were able to take advantage of the situation . . . to secure a real improvement . . . and to accomplish this . . . when *the total wealth of the community as a whole had suffered a decrease* . . . [italics added] Thus, the working classes improved their *relative* position in the years following the war, as against all other classes, except that of the profiteers . . . Was it due to a permanent modification of the economic factors . . . ? Or was it due to some temporary and exhaustible influence connected with inflation . . . ?" [pp. 25–27]

Two similarities between my approach and that of Keynes are evident: (1) inflationary processes are related to periods in which the total wealth of the community has diminished and the distribution of wealth has been perceived to be altered (for example, during and after wars), and (2) the goal he seems to attribute to the working classes is that of improving their relative position in the distribution of wealth.

Is the process Keynes described relevant to today's economy? The answer seems to be positive: while it is true that the problems today might not be related to diminished wealth, the model presented here shows that exactly the same processes may occur when expectations are not realized. In these circumstances, several sectors may put pressures on the government, and phenomena popularly labeled as "cost-push" inflations, "catch-up of wages" among sectors, or "wage-price inflationary spirals" may result (with the cooperation of central banks). But there is a profound difference between this view of the "cost-push" inflationary process and the traditional, much criticized one.[3] In the traditional "cost-push" analysis, it is never made clear why employees, unions, or governments suddenly change their behavior, and why greater political pressures are suddenly put on governments. Thus the whole point of departure for such "cost-push" models of inflation is ad hoc. In contrast, in the model presented in chapter 1 and applied here, the change in behavior is explained and one can predict the circumstances in which the probability that inflations will occur increases: they are linked with a perception of leapfrogging. Then the demands for "higher wages" (i.e., for maintaining wages at a promised level that "the market" can no longer maintain, but political pressures may) and for government intervention represent a demand for a redistribution of the pie.

Political Pressures and the National Debt

An alternative view of the inflationary process is presented in Keynes's (1923) *Tract on Monetary Reform*, in which inflation is viewed as a tax that can diminish the government's deficits by reducing the burden of outstanding liabilities fixed in nominal terms. As Keynes points out, however,

It would be too cynical to suppose that in order to secure [these] advantages . . . , governments depreciate their currency *on purpose*. As a rule they are, or consider themselves to be, driven to it by their necessities. The requirements of the Treasury to meet sudden exceptional outgoings—for a war or to pay the consequences of defeat—are likely to be the original occasions of, at least *temporary*, inflation. But the most cogent reason for *permanent* depreciation, that is to say *devaluation*, or the policy of fixing the value of currency permanently at the low level to which a temporary emergency has driven it, is generally to be found in the fact that a restoration of the currency to its former value would raise the recurrent annual burden of the fixed charges of the National Debt to an insupportable level. [p. 53, italics in original]

What does Keynes's "insupportable level" mean? As his arguments and numerical example point out, these words refer to the distribution of wealth if the inflationary strategy was not pursued:

> Experience shows with great certainty that the active part of the community will not submit in the long run to pay too much to vested interest, and, if the necessary adjustment is not made in one way, it will be made in another—probably by the depreciation of the currency . . . At the end of 1922, the internal debt of France, excluding altogether her external debt, exceeded 250 milliards francs . . . The total normal receipts under the provisional budget for 1923 are estimated at around 23 milliards. That is to say, the service of debt will shortly absorb, at the value of the franc current early in 1923, almost the entire yield of taxation. Since other government expenditures in the ordinary budget cannot be put below 12 milliards a year, it follows that . . . the yield of taxation must be increased permanently by 30 per cent to make both ends meet. If, however, the franc were to depreciate to (say) 100 to the pound sterling, the ordinary buget could be balanced, by taking little more of the real income of the country than in 1922. [pp. 58–59]

How relevant is this argument for understanding more recent inflationary processes?

In spite of the fact that neither wars nor defeats can explain the current problems with either the internal debt or the deficits, there have been other unexpected events that have led to problems similar to those described by Keynes. Consider the currently debated issue of social-security benefits: some economists say that these benefits have been, inadvertently, linked to indexes that have increased by 12% to 23% more than alternative, better indexes (the numbers depend on the economist and his calculations). In addition, although people's average age at death has increased (perhaps unexpectedly), retirement age has not changed, and the number of recipients of social-security benefits has therefore increased. If social-security benefits are to be maintained at their promised level without changing the retirement age, problems similar to those described by Keynes may emerge (or may already have done so), since taxes on wages or increased government borrowing must pay for the unexpectedly increased transfer payments. In addition, other unexpected changes have occurred in recent years, such as the rise of import prices (oil, in particular), which has diminished the wealth of the importing countries and which could have led as well to problems similar to those raised by Keynes. Still, in other countries with prolonged inflation (Israel, for example), Keynes's arguments seem to remain relevant.[4] Thus Keynes's description may be applicable to all situations in which the government's internal debt has increased relative to the total income in the economy. Is there no escape, then, from inflationary policies in such circumstances?

Unfortunately, the answer is that it is easier to say what government *cannot* allow itself to do than what it *should* do if the unexpected events described above take place. The government will have to *change* its policies and either redistribute wealth or promise some future redistribution of benefits, and it *cannot* stick to its traditional strategies. The rules must be changed not because somebody in the government randomly decides to do so, but because unexpected events have taken place, and the economy (including the government sector) must adapt to them.

But it is not clear what governments *should* do in such circumstances. Keynes (1923) argues that redistributing wealth and diminishing the internal debt by a "capital levy" is preferable to inflation, since such a tax would be imposed less randomly than the inflationary tax. However, Keynes admits that for a variety of reasons inflation may turn out to be politically more attractive, and he points out that "The power of taxation by currency depreciation is one which has been inherent in the state since Rome discovered it. The creation of legal tender has been and is a government's ultimate reserve; and no state or government is likely to decree its own bankruptcy or its own downfall so long as this instrument still lies at hand unused" (p. 64). However, Keynes fails to mention that the decision to gamble on inflationary policies may not be in the interest of a minor ruling political group alone, but may be perceived as attractive by the majority, considering the alternatives of the costly ways of raising money and the length of time that may elapse before it is raised. Indeed, while today's prolonged inflationary periods are perceived by some as unjust and costly, in earlier times reactions to them were frequently favorable.

Despaux's (1936) massive and very detailed chronological summary of inflationary episodes throughout history suggests that in cases of urgency (wars, threats of social unrest, etc.), inflation was viewed as a just and efficient method both for raising taxes and redistributing wealth. The reason that this view seems to be missing from today's economic literature on inflation may be that economists have arbitrarily separated monetary theory from theories of wealth redistribution and have concentrated on a technical feature of inflation, namely, the correlation between changes in the money supply and changes in price levels, thus putting all inflationary episodes in the same basket—more will be said on this subject below. Next, an examination of Keynes's recommendations for an inflationary policy in the *General Theory* leads to two conclusions: first, that it may not be very useful to put all inflationary episodes in the same basket, and second, that one cannot rely *at all* on the policies derived from the conventional interpretations of the *General Theory* in order to shed light on any of our problems today.

2. Inflation, Keynes, and Keynesian Policies

The model presented in chapter 1 has suggested that one should look for inflationary periods not randomly, but during times when the distribution of wealth has been unexpectedly altered, or when the realized wealth of a community was less than the one expected. The cases that have been most frequently discussed in the economic literature do indeed refer to such circumstances: both Keynes's and Friedman's analyses of inflations referred either to postwar periods or to the famous hyperinflations of Europe between the two world wars (periods that have been correlated with large increases in the internal and external debts of the countries concerned).[5]

Yet, if one opens some standard "macroeconomic" textbooks, one will find that inflations (and monetary expansions) have, for some economists, positive connotations: these economists claim to rely on Keynes's *General Theory*, published in 1936, when they associate inflations in general with rising employment and rising output. If such theories find currency (for a while) one may observe inflationary periods that will be associated not necessarily with conditions such as I have discussed, but with circumstances in which politicians gamble on the views of these economists. In such circumstances inflation introduces additional uncertainty into the economy and has harmful effects.

Let us, then, make the following points in this section: Keynes's advocacy of "inflation" in the *General Theory* seems reasonable for the times when he wrote the book, and his views do seem related to those presented in chapter 1. These arguments will help us to understand why some Keynesian policies might have worked if they had been implemented then, and why economists looking at their success attributed them to the wrong models and deduced that Keynesian inflationary policies fitted *all* circumstances.[6] The extensive quotations below from the *General Theory* serve to document my interpretation of the work, which is different from the customary one.

Keynes starts the *General Theory* by refuting the following two postulates of the classical theory:

> I *The wage is equal to the marginal product of labor . . .* II *The utility of the wage, when a given volume of labor is employed, is equal to the marginal disutility of that amount of employment.* [p. 5; italics in original]

But Keynes's argument against the classical school is *not* based on some rigidity of wages due to the existence of labor unions, as we see in the following statement:

> The classical school reconcile this phenomenon [that more labor would be forthcoming at the existing money wage if it were demanded] . . . by arguing that . . . this situation is due to an open or tacit agreement amongst workers not to work for less . . . If this is the case such unemployment, though apparently involuntary, is not strictly so, and ought to be included under the category of "voluntary" unemployment. . . ." [pp. 7–8]

This quotation clearly indicates that the definition of "involuntary unemployed" has nothing to do with unions; Keynes considers the existence of unions or of contracts, or, in general, the "effects of collective bargaining," as consistent with optimizing, unconstrained behavior. But his view of the behavior of the labor market is different:

> Now ordinary experience tells us, beyond doubt, that a situation where labour stipulates (within limits) for a money-wage rather than a real wage, so far from being a mere possibility, is the normal case. Whilst workers will usually resist a reduction of money wages, it is not their practice to withdraw their labour whenever there is a rise in the price of wage-goods. [p. 9]

The meaning is clear: the amount of labor supplied is different when the decrease in real wages is due to a decrease in the money wage, the price level staying constant, rather than an increase in the price level, the money wages staying constant.[7] This empirical evidence[8] indicates, in Keynes's opinion, that the supply of labor is not only a function of the real wage. The interpretation generally given to this passage and other similar ones in the *General Theory* is that of "money illusion." However, it is hard to understand how such an interpretation has gained currency,[9] since Keynes explains several times what the rationale for such asymmetrical behavior is:

> Since there is imperfect mobility of labor, and wages do not tend to an exact equality of net advantage in different occupations, any individual or group of individuals, who consent to a reduction of money wages relatively to others, will suffer a *relative* reduction in real wages, which is a sufficient justification for them to resist it. [p. 14, italics in original]

We shall see how the assumptions of relative behavior and of struggles over the distribution of wealth explain this quotation and the meaning of "involuntary unemployment." But first here are some additional quotations from the *General Theory* on which my arguments will rely.

When describing the behavior of workers faced by an increase in the price level Keynes says:

> . . . it would be impracticable to resist every reduction of real wages, due to a change in the purchasing-power of money which affects all workers alike; and in fact reductions of real wages arising in this way are not, as a rule, resisted unless they proceed to an extreme degree. . . . In other words, the struggle about money wages primarily affects the *distribution* of the aggregate real wage between different labour-groups. [p. 14, italics in original]

Here we see that the jealousy and the competition among groups, and not only real wages, have been assumed to affect behavior in the labor market.[10]

When Keynes concludes the policy implications of his model with respect to the labor market in chapter 19, he writes:

> Except in a socialized community . . . there is no means of securing uniform wage reductions for every class of labor. The result can only

be brought about by a series of gradual, irregular changes, justifiable on no criterion of social justice or economic expediency, and probably completed only after wasteful and disastrous struggles. . ." [p. 267]

The behavior pattern described in these as well as in other statements in the book (especially in chapter 19), clarifies the definition of "involuntary unemployment" and leads to Keynes's recommendations of inflationary policies. Let us show that both rely on the simple idea that "everybody is sitting in the same boat," an idea that implies that people do make comparisons with others when they are shaping their behavior.

One may attribute to an individual in Keynes's world the following attitude: why should an unemployed person with given skills accept a wage that is lower than the actual wage earned by somebody with similar skills? This would be "unfair" and "unjust." If indeed such a concept of "justice" or "fairness" shapes people's attitudes, then the definition of "involuntary unemployment" becomes clear: those actually employed do not expect somebody looking for a job to accept a real wage lower than the one they are currently earning. Therefore, when somebody is offered a wage lower than the actual real wage and does not accept it, he is viewed by Keynes as "involuntarily unemployed." Clearly, this definition cannot be derived from the neoclassical approach, which would consider the individual's behavior in not accepting the job to be a voluntary act. But this definition holds if the notion of justice just laid out is shared by the majority.

From here it is only a short step to Keynes's other assumption on the behavior of the labor market, and to his policy implications. The behavior of a group of workers as reflected in the previous quotations is characterized by the following attitude: why should they suddenly consent to a *relative* reduction in their wages?[11] This would be "unjust." However, if everybody's wage is reduced uniformly, *relative* wages will remain unchanged and employment will increase.[12] This happens because somebody who was previously "involuntarily unemployed" is expected now to accept the lower real wage, since the wages of those employed have decreased. These are the reasons in Keynes's view for "just" persons to accept *inflation* as a means by which both the national income equilibrium *and* the equilibrium distribution of shares are restored.[13] To quote:

. . . it is fortunate that the workers . . . inasmuch as they resist reductions of money wages, which are seldom or never of an all-around character, . . . do not resist reductions of real wages, which are associated with increases in aggregate employment and leave relative money-wages unchanged . . ." [p. 14]

This fairness is the reason for the asymmetrical behavior toward the source of the change in the real wage: a reduction in real wages due to a change in the purchasing power of money "affects all workers alike" (or "secures a uniform wage reduction," p. 267), and will therefore be accepted exactly because of the concept of "justice" workers have in mind. In simple words, this notion of justice is that "everybody is sitting in the same (temporarily sinking) boat."[14] (Notice the similarity between Keynes's view and the model presented in the first chapter: Keynes was examining a situation in which the community's wealth has unexpectedly diminished.)

It is evident from Keynes's correspondence during the preparation of the *General Theory* and after its publication that in spite of the fact that the definition of "involuntary unemployment" was never understood by other economists,[15] Keynes never revised his definition, although he promised to do so.[16] Only Harrod seemed to understand Keynes's view of the behavior of the labor market in the way presented here:

> . . . if there is no strike against rising prices, that may be because it is thought (even if wrongly) that the whole community (including capitalists) is in the same boat. [1973a, p. 528]

But Harrod did not agree with this view and wrote "that the industrial struggles are almost always intended to strike at capitalists, not fellow workers" (p. 528). He recommended that Keynes avoid this point. Keynes never answered on this issue and left the paragraphs unchanged.

But if this was Keynes's view of the behavior of labor markets, one can give a simple summary of the inflationary process that Keynes described in the *General Theory*: there is a given distribution of incomes among the different competing groups in the economy. For an unexpected, exogenous reason expectations change,[17] the demand for investment (and thus for labor) decreases, and the real output and the price level fall. Since wages are contractually fixed in nominal terms—and the importance of this point will be explained next—"the struggle for the distribution of aggregate real wage" begins. Suppose now that the price level is increased, bringing a uniform decrease in real wages. Since the groups of workers are now interested in their relative share, further struggles will be avoided, "involuntary unemployment" disappears, the economy returns to the original national income equilibrium, and the equilibrium distribution of shares is restored.[18]

Let us elaborate Keynes's description of this process and examine in more detail the relationship that he assumed existed between money wages, inflation, and relative behavior. Keynes had one further objection to the classical postulates:

> To sum up: there are two objections to the second postulate of the classical theory. The first relates to the actual behaviour of labour . . . But the other . . . flows from our disputing the assumption that the general level of real wages is directly determined by the character of the wage bargain. . . The assumption that the general level of real wages depends on the money-wage bargains between the employers and the workers is not obviously true. [1936, pp. 12–13]

In other words, although wage contracts between employees and employers are determined in terms of *money*, the real wage may fluctuate. Keynes assumes that a money wage bargain is struck, the wages defined by the bargain remain constant in nominal terms until future negotiations, and this waiting time defines the short-run.[19] The existence of contracts in nominal terms sheds light on the arguments and policy implications presented above. The fact that money wages are constant for a while explains Keynes's assertions that the individual or groups of laborers possess information on the nominal value of these wages, and that a uniform decrease in real wages can be obtained through a gradual increase in the price level. In the absence of money-wage contracts, neither the information on the level of wages nor the "uniformity" in the decrease of real wages could hold.

In these arguments Keynes tried to integrate money with value theory and to capture the essence of a monetary economy. A monetary economy must be characterized not merely by the existence of money but by the existence of nominal contracts. If nominal contracts ceased to exist, prices would fluctuate wildly, money would not fulfill its normal purposes, and the economy would cease to be a monetary one. This leads to Keynes's view that money wages "should be inflexible in terms of money" and that money-wage flexibility—a policy recommended by other economists at that time—would only make the situation worse. This interpretation, which was also Lerner's (1952),[20] explains the following statements:

... The attribution of relative stability to real wages is . . . a mistake in fact and experience. If, indeed, some attempt were made to stabilize real wages by fixing wages in terms of wage goods, the effect could only cause a violent oscillation of money prices. . . . That money-wages should be more stable than real wages is a condition of the system possessing inherent stability. [1936, p. 239]

If money wages are inflexible . . . the greatest practical fairness will be maintained between labour and the factors whose remuneration is contractually fixed in terms of money. . . . If important classes are to have their remuneration fixed in terms of money . . . social justice and social expediency are best served if the remunerations of *all* factors are somewhat inflexible in terms of money. . . . [Therefore] it can only be an unjust person who would prefer a flexible wage policy to a flexible money policy. . . . [p. 268; italics in original]

Much earlier, Keynes made some similar points in "An Economic Analysis of Unemployment" (lecture 3: "The Road to Recovery" [1931]):

... the . . . reason for wishing prices to rise [back to parity with what a few months ago we considered the established levels of our salaries, wages] . . . is on grounds of social justice and expediency which have regard to the burden of indebtedness fixed in terms of money. If we reach a new equilibrium by lowering the level of salaries and wages, we increase proportionately the burden of monetary indebtedness.

In doing this we should be striking at the sanctity of contract. [1973a, pp. 360–61]

Although Keynes mixes normative and positive statements, one may conclude that his objections against fluctuations in the price level arose because he realized that long-term monetary contracts, rather than just "money," characterize a monetary economy.[21] These views suggest why Keynes thought that in a monetary economy money-wage flexibility does not provide a solution for returning to equilibrium, a point much stressed in the *General Theory*:

> There may exist no expedient by which labour . . . can reduce its *real* wage to a given figure by making revised *money* bargains with the entrepreneurs. This will be our contention. [1936, p. 13; italics in original]

Keynes's explanation was that if money wages drop, one cannot assume other things to stay constant, since with the drop in money wages, the expected price level will also drop. And the lower expected price level implies that although money wages have dropped the real wages employers expect to pay have not dropped. Therefore, flexible money wages will not restore equilibrium; moreover, since flexible money wages lead to a flexible expected price level, the economy will cease to be a monetary one.

In conclusion: the "inflationary policy" Keynes recommended referred to a situation in which the price level has first *unexpectedly diminished*. Perhaps much heated discussion on the *General Theory* could have been avoided if Keynes had used the word "reflationary" instead of "inflationary", or if economists had paid more attention to the fact that Keynes defined the word "inflation" *differently* from the way in which his contemporaries did (and from the way in which we define it today), as this quotation from the *General Theory* shows:

> The view that *any* increase in the quantity of money is inflationary (unless we mean by *inflationary* merely that prices are rising) is bound up with the underlying assumption of the classical theory that we are *always* in a condition where a reduction in the real rewards of the factors of production will lead to a curtailment in their supply . . . [p. 304; italics in original]

The observation that Keynes casually altered the meaning of words is by no means novel. One of his contemporaries, the historian Marc Bloch (1953), wrote in his notes, published after his death, that "there is hardly one of [Keynes's] books in which he does not, from the beginning expropriate terms, usually pretty well established, in order to decree entirely new meanings for them, meanings which sometimes vary from work to work, but, in any case, intentionally depart from common usage" (p. 176). The danger of such misuse is that such words frequently continue to misrepresent the ideas of their author for several generations.

Since for readers accustomed to Keynesian textbook models, my reading of his views may come as a surprise, it is useful to note that at first Keynes's policy implications did not make a great impression on his American contemporaries, because they did not find anything new in them and interpreted his views as I did above:

> It comes as a great surprise . . . that no American economist between 1929 and 1936 advocated a policy of wage-cutting; the leaders of the American profession strongly supported a programme of public works and specifically attacked the shibboleth of a balanced budget. A long list of names including Slichter, Taussig, Schultz, Yntema, Simons, Gayer, Knight, Viner, Douglas and Y. M. Clark concentrated mainly at the Universities of Chicago and Columbia, . . . declared themselves in print well before 1936 in favour of policies that we would today call "Keynesian." . . . Orthodox economists had no difficulty in explaining the persistence of unemployment. The government budget in both the United States and Britain was in surplus during most years in the 1930s. It did not need Keynes to tell economists that this was deflationary. [Blaug 1976, pp. 162–63]

Ayres (1944) elaborates the suggested policy:

> In [the] event (that the total amount of income . . . somehow fails to flow into the market . . .) the recreation of money values by deficit financing up to the amount of purchasing power necessary to absorb the product of industry at full employment would only be a salvage of purchasing power already lost by its former owners and so the whole community. It is this circumstance which has led to the recognition of deficit financing by many respectable economists in recent years . . . Indeed there is no good reason why such a program should not take the form of the outright issue of currency in the amount so indicated. [p. 276]

Neither did Keynes's view on long-term monetary contracts strike a cord; Irving Fisher, one of the fathers of "monetarism," had paid much attention to them long before Keynes.[22] Their views seem close, as the following quotations from Fisher's works show:[23]

> [The] monetary yardstick is used for long-term contracts, in which the dollars of today are exchanged for dollars of the future. If a man contracts to pay you a certain number of dollars today, it makes a tremendous difference both to him and to you whether the dollars in which you are paid have meanwhile shrunk or expanded. It makes a tremendous difference, for instance, to a bondholder. [1928, p. 59]

> Salaries and wage contracts are also like bonds. Although they run for shorter periods, and may be readjusted, they seldom are readjusted promptly or fully. [p. 83]

Fisher connects the existence of these contracts with changes in the price level as follows:

. . . salaries . . . do not rise or fall with inflation or deflation as [the prices of] commodities do. . . . [They] are kept stationary by contract. . . . This later case is what interests us in a depression: the pinching of the business man's income between the upper millstone of lowering commodity prices and the lower millstone of fixed or semi-fixed expenses, including the interest on his debts. [1934a, p. 36]

The mechanism Fisher has in mind is this: since debts and wages are fixed in nominal terms, a decrease in the price level decreases nominal revenues (although not necessarily the real ones), while leaving some costs unchanged. Thus the probability of a bankruptcy is increased. Since according to Fisher bankruptcy imposes a dead-weight loss on the society, deflation causes a decrease in wealth.[24] His solution for avoiding this decrease was similar to Keynes's:

On the other hand, if the foregoing analysis [the debt-deflation theory] is correct, it is always economically possible to stop or prevent such a depression simply by reflating the price level up to the average level at which outstanding debts were contracted by existing debtors and assumed by existing creditors, and then maintaining that level unchanged. [Fisher, 1933b, p. 346]

While Fisher emphasizes the role of long-term monetary contracts in financial markets, and Keynes in labor markets, their views seem close (the disagreement in their theories is related to the source of the initial decrease in the price level).

If this interpretation of Keynes's policies is correct, one can understand why, if they are implemented, they may have the desired consequences in the very restricted circumstances identified above. What is more difficult to understand is how the simple algebra of the macroeconomic, so-called IS-LM models, or the mathematically more sophisticated ones of Benassy (1975), Grandmont and Laroque (1976), and others are viewed as being related to the *General Theory*. It is true that the IS-LM models were based on Hicks's "Mr. Keynes and the Classics,"[25] but neo-Keynesians did not pay any attention to Hicks's reservations:

Admittedly, it follows from this theory that you may be able to increase employment by direct inflation: but whether or not you decide to favour that policy still depends upon your judgement about the probable reaction on wages. [1937, p. 150]

It is clear, then, why Keynes accepted this model: Hicks left both the issue of the behavior of wages and the definition of unemployment open.[26] Advocates of the IS-LM models later seemed to neglect even Hicks's reservations[27] and did not pay much attention to the definitions Keynes started from. Instead, they followed the classical approach of exogenous wage rigidity which, as shown, was *not* Keynes's departure point.

Policy Implications

Some of Keynes's policy recommendations in the *General Theory* fit only those circumstances in which, unexpectedly, the price level turns out to be significantly less than expected. In these circumstances increased government expenditures and deficits through a one-shot *monetary expansion*—even for digging holes in the ground (as Keynes put it)—may improve the situation, since they restore the value of real wages and debts to their expected level. This policy recommendation makes sense because Keynes assumes that workers react to changes in their relative position in the distribution of wealth and that nominal wages are not instantaneously revised downward when price levels unexpectedly fall. But it is useful to reemphasize that at that time many economists, monetarists in particular, recommended exactly the same policies as Keynes without making these assumptions. Instead they assumed that such a policy of monetary expansion could restore the value of real debt in the private sector to its expected value and thus prevent further bankruptcies.

While Fisher's and Keynes's policy recommendations were similar, their views of the source of the disturbance were not: according to Keynes they were due to fluctuations in the "animal spirits" (that, according to him, possess ordinary people, but not those working for the government), while according to Fisher they were due to mistakes in monetary policy. This different perception of the source of the trouble led to different *long-term* policy implications: an active role for policy makers for Keynes, and a passive one for Fisher, since according to the latter the government can, at best, only correct its own previous mistakes (and can do this when no crowding out occurs). But in the short run, the policies they recommended were, in some sense, similar (more about Keynes's views later).

Can these viewpoints shed light on discussions on monetary expansion and government expenditures today? The answer is: hardly. For there is not much evidence that inflation rates have turned out to be *significantly lower* than expected (as they were during the Great Depression when price levels dropped by 30% during a relatively short period of time). Thus the circumstances identified until now (circumstances in which government intervention is helpful) may not be of much practical relevance today. However, the circumstances identified next may have such relevance.

Uncertainty, Perceptions, and Keynes's Views

Far be it from me to pretend that the previous discussion provides a summary of Keynes's views. The arguments above present just *one* view of the *General Theory*, a "theory" which, one must admit, is neither general nor "a theory," but includes many theories. Indeed, there is agreement among economists that what Keynes intended to present in his book was not a theory, but a view of the economy where disequilibrium persists, and where risk, uncertainty, and expectations play the central roles.[28] Since these notions were not rigorously introduced in theories dealing with what we call today "macroeconomic" variables, Keynes felt forced first to adapt the economists' vocabulary to his views. That was a mistake: it led to obscurity, and even today leads to much confusion and controversy. For Keynes was wrong even to make the attempt. Words that can be defined in models where uncertainty is formally introduced have no meaning in models of "certainty"—"confidence" and "animal spirits" have no roles in the latter. And the vocabulary of the economists, which was invented to form definitions in models of "certainty," loses its original meaning once uncertainty is introduced.

In the chapters, sections, and paragraphs where Keynes *abandons* these attempts and uses ordinary language, his arguments become quite clear. Yet it is exactly these lucid parts of the *General Theory* that most economists have avoided discussing. The next section provides the reason: Keynes's statements in these lucid parts rely on a view of human nature that had no foundations in economic analysis and indeed cannot be incorporated within the traditional framework. So let us first remind ourselves of some of Keynes's straightforward arguments, expressed in clear, ordinary language, and then see how they fit in with my views.

Keynes's Views on Confidence, Spending, and Animal Spirits

According to Keynes (1936):

> . . . the traditional analysis is faulty because it has failed to isolate correctly the independent variables of the system. Saving and Investment are the determinates of the system, not the determinants. They are the twin results of the system's determinants, namely, the propensity to consume, the schedule of the marginal efficiency of capital and the rate of interest . . . [pp. 183–84]

But Keynes emphasizes that the first two depend in fact on psychological factors and habits:

> The amount that the community spends on consumption obviously depends (i) partly on the amount of its income, (ii) partly on the other objective . . . circumstances, and (iii) partly on the subjective needs and the psychological propensities and habits of the individuals composing it . . . [pp. 90–91].

> The *state of confidence*, as they term it, is a matter to which practical men always pay the closest and most anxious attention. But economists have not analyzed it carefully and have been content, as a rule, to discuss it in general terms. In particular it has not been made clear that its relevance to economic problems comes in through its important influence on the schedule of the marginal efficiency of capital . . . There is, however, not much to be said about the state of confidence *a*

priori. Our conclusions must mainly depend upon the actual observation of markets and business psychology. [pp. 148–49; italics in original]

Thus toward the end of the book he concludes that

. . . we can sometimes regard our ultimate independent variables as consisting of . . . the three fundamental psychological factors, namely the psychological propensity to consume, the psychological attitude to liquidity and the psychological expectation of future yield from capital-assets . . . [p. 246]

Unsurprisingly, therefore, fluctuations in these three *psychological* factors are viewed by Keynes as determining the level of employment in an economy, as the following statements suggest:

. . . every weakening in the propensity to consume regarded as a *permanent habit* must weaken the demand for capital as well as the demand for consumption. [p. 106, italics added]

Thus if the animal spirits are dimmed and the spontaneous optimism falters, leaving us to depend on nothing but a mathematical expectation, enterprise will fade and die . . . . [p. 162]

The model presented in chapter 1 and the discussion in chapter 2 have provided the mathematical proofs for the latter statement. As to investment and risk taking, Keynes wrote:

But individual initiative will only be adequate when reasonable calculation is supplemented and supported by animal spirits, so that the thought of ultimate loss which often overtakes pioneers . . . is put aside as a healthy man puts aside the expectation of death. [p. 162]

We are . . . reminding ourselves that human decisions affecting the future, whether personal or political or economic, cannot depend on strict mathematical expectation, since the basis for making such calculations does not exist; and that it is our innate urge to activity which makes the wheels go round, our rational selves choosing between the alternatives as best as we are able, calculating where we can, but often falling back for our motive on whim or sentiment or chance. [pp. 162–63]

These statements, too, have been given precision within my model.

What, then, is causing, according to Keynes, a "crisis" of increased unemployment and lowered economic activity? As noted above, a permanent weakening of the propensity to consume can lead to such an outcome, although according to Keynes

. . . a more typical, and often the predominant, explanation for the crisis is, not primarily a rise in the rate of interest, but a *sudden* collapse in the marginal efficiency of capital. [p. 315, italics added]

And, once this occurs,

> . . . it is not easy to revive the marginal efficiency of capital, determined as it is, by the uncontrollable and disobedient psychology of the business world. It is the *return of confidence* to speak in ordinary language, which is so insusceptible to control in an economy of individualistic capitalism. This is the aspect of the slump which bankers and businessmen have been right in emphasizing, and which the economists who have put their faith in a "purely monetary" remedy have underestimated. [p. 317; italics added]

These views have led Keynes to his conclusions on government expenditures, "the paradox of savings," and the role of government:

> . . . the economic principle, on which the practical advise of economists has been almost invariably based, has assumed, in effect, that, *cet. par.*, a decrease in spending will tend to lower the rate of interest and an increase in investment to raise it. But if what these two quantities determine is, not the rate of interest, but the aggregate volume of employment, then our outlook on the mechanism of the economic system will be profoundly changed. A decreased readiness to spend will be looked on in quite a different light if, instead of being regarded as a factor which will, *cet. par.*, increase investment, it is seen as a factor which will, *cet. par.*, diminish employment. [p. 185; italics in original]

These arguments seem straightforward: when people become more pessimistic and lower their aspirations for the future, they both invest less and spend less, lowering the level of employment and diminishing what we measure as economic activity. Can the government do something in these circumstances? Keynes's answer is positive, although by no means related to the simplistic IS-LM models: according to him decision makers within governments *may* restore confidence by increasing expenditures, but not just by pursuing a monetary policy. How and in exactly what circumstances will be discussed after reinterpreting Keynes's views in the light of the model.

Keynes's Vocabulary Reinterpreted

1. Keynes's intention was to examine situations of *disequilibrium*. The model on which all the arguments in this book rely is a precise disequilibrium model which starts by looking at people's motivations. It suggests that when expectations and aspirations are realized, people avoid some risky courses of action. But when one's wealth suddenly diminishes relative to that of others, then one starts betting on new ideas. Within the model this continuous betting process constitutes a "disequilibrium." Once stability is achieved, and a new order is found, the betting on ideas stops.

2. Keynes has pointed out clearly that since the ultimate independent

factors are "psychological" (in particular those concerning investments), one must therefore start the examination by looking at the motivations of individuals—my views include these features as well.

3. Keynes emphasized the central role of "animal spirits" in understanding fluctuations in measured economic activity, rather than the role of mathematical expectations. My views give precision to these terms: the creative elements of "thinking" (i.e., of betting on ideas) lie within the emotions, and the probability that appears in the model when the entrepreneurial act is defined represents a subjective judgment. Also, Keynes's statement that when "animal spirits are dimmed . . . enterprise will fade and die" (p. 162) holds true within the model. For when aspirations are realized, the animal spirits are latent, and no bet on new ideas is made (this concept can be found in Mandeville's statements as well as in the works of a number of other writers).

4. Keynes always emphasized (together with Knight 1921) the distinction between "risk" (or "mathematical expectations") and "uncertainty," arguing that the latter is related to the flourishing of business enterprises and to profits. These views are accurate: new ideas, if they turn out succesfully, will be judged by an outsider as representing an increase in "profits" (at the same time that the frustration of individuals who fell behind, and the emotional toll that results, will not be measured). But when no such bets on new ideas take place, there will be no profits, the economy will be "stationary," and mathematical expectations will become relevant.[29]

5. "The *paradox of savings*." Let us assume that W_1 represents the wealth one expects to achieve, but $W_0 < W_1$ is the wealth one owns at a point in time, which an outsider may measure (W_1 "exists" only in the individual's mind). The same outsider may measure C as his level of consumption and define C/W_0 as his average propensity to consume. But C depends on W_1, and the proportion C/W_0 is greater than C/W_1. If realization start to be frustrated and expectations are lowered (through the mechanism suggested by Simon 1959) to, let's say, W_0, one's consumption will be adjusted, too, and will now depend on W_0 only (when as you will recall, $C(W_1) > C(W_0)$). The economy will settle at a *lower* level of measured economic activity, since some people will stop betting on new ideas. Thus, indeed, measured savings will now have increased (since $W_0 - C(W_1) < W_0 - C(W_0)$) and the level of economic activity will be lowered, since people exert themselves less. Keynes gives a *negative* connotation to this outcome (but as shown below Mandeville does not; he sees that it has benefits, too: the smaller emotional toll paid by people exerting themselves less). But in the *General Theory* this paradox has to do with frustration, discouragement, and the lowering of aspirations, rather than with the assumptions on interest rates and the mechanism postulated by the IS-LM models. (Interest rates represent just one particular relative price, the effect of which is already implicitly incorporated in one's perception of "wealth.") It is this process of *preventing*

aspirations from being lowered that preoccupied Keynes and led to his view that monetary policy could not provide remedies, while some direct government expenditures could—we shall later see how.

Was Keynes's view of the paradox of savings novel? The following, sometimes extensive quotations show that Keynes himself linked his views with those of earlier writers, with Mandeville's *Fable of the Bees* in particular. It is useful to point out these similarities since they lead to a better understanding of Keynes's complex views and especially illuminate his opinions on the issue of government intervention.

Keynes and the Fable of the Bees

When Keynes discusses the implications of his views in the final chapters of the *General Theory* he mentions several other writers who in his view have held views similar to his. Among them is Malthus, who wrote in the preface to *Principles of Political Economy*:

> Adam Smith has stated that capitals are increased by parsimony, that every frugal man is a public benefactor, and that the increase of wealth depends upon the balance of produce over consumption. That these propositions are true to a great extent is perfectly unquestionable . . . But it is quite obvious that they are not true to an indefinite extent, and that the principles of saving, pushed to excess, would destroy the motive to production. *If every person were satisfied with the simplest food, the poorest clothing, and the meanest houses, it is certain that no other sort of food, clothing, and lodging would be in existence* . . . The two extremes are obvious; and it follows that there must be some intermediate point, though the resources of political economy may not be able to ascertain it, where, taking into consideration both the power to produce and the will to consume, the encouragement to the increase of wealth is the greatest. [quoted by Keynes 1936, p. 363; italics added]

Keynes also quotes extensively from Gesell's and Hobson's discussions on these and related subjects and concludes that "Mandeville, Malthus, Gesell and Hobson . . . following their intuitions, have preferred to see the truth obscurely and imperfectly rather than maintain error, reached indeed with clearness and consistency and by easy logic but on hypotheses inappropriate to the facts" (p. 371) (here Keynes implicitly puts himself in the Mandeville-Malthus category).

But it is the similarity with Mandeville's views, which Keynes quotes at length in the *General Theory*, that can best illustrate his source of inspiration. So let us present Mandeville's views at some length.

Mandeville (1932) satirizes the conflict between virtue and prosperity. One day the bees are smitten with virtue and begin to lead sober lives, abandoning pomp and pride and adopting frugal, modest ways. The outcome is the Great (Bee) Crash:

But, Oh ye Gods! What Consternation,
How vast and sudden was th' Alteration!
In half an Hour, the Nation round,
Meat fell a Peny in the Pound. [p. 28]

As Pride and Luxury decrease,
So by degrees they leave the Seas.
Not Merchants now, but Companies
Remove whole Manufactories.
All Arts and Crafts neglected lie;
Content, the Bane of Industry,
Makes 'em admire their homely Store,
And neither seek nor covet more. [pp. 34–35]

The Shew is gone, it thins apace;
and looks with quite another Face.
For 'twas not only that They went,
By whom vast Sums were Yearly spent;
But Multitudes that liv'd on them,
Were daily forc'd to do the same.
In vain to other Trades they'd fly;
All were o'er-stock'd accordingly. [p. 32]

The Price of Land and Houses falls;
Mirac'lous Palaces, whose Walls,
Like those of Thebes, were rais'd by Play,
Are to be let; . . . [p. 32]

The building Trade is quite destroy'd;
Artificers are not employ'd;
No Limner for his Art is fam'd,
Stone-cutters, Carvers are not nam'd. [p. 33]

Before people lowered their expectations, however, "suddenly taking into their heads to abandon luxurious living, and *the State to cut down armaments*"[30] (Keynes 1936, p. 360; italics added), the situation was quite different:

The Root of Evil, Avarice,
That damn'd ill-natur'd baneful Vice,
Was Slave to Prodigality,
That noble Sin; whilst Luxury
Employ'd a Million of the Poor,
And odious Pride a Million more:
Envy itself, and Vanity,
Were Ministers of Industry;
Their darling Folly, Fickleness,
In Diet, Furniture and Dress,
That strange ridic'lous Vice, was made
The very Wheel that turn'd the Trade.

[Mandeville 1932, p. 32; italics added]

> Thus every Part was full of Vice,
> Yet the whole Mass a Paradise. [p. 24]

The similarity with the implications of the formal model in chapter 1 is obvious. Mandeville leaves no doubt in his long remarks on the meaning and implications of these passages:

> As this prudent Oeconomy, which some People call / Saving, is in private Families the most certain Method to increase an Estate, so some imagine that whether a Country be barren or fruitful, the same Method, if generally pursued (which they think practicable) will have the same Effect upon a whole Nation, and that, for Example, the English might be much richer than they are, if they would be as frugal as some of their Neighbours. This, I think, is an Error. [p. 182]

Sounds familiar? Mandeville's explanation for this "paradox of savings" is clear: when people lower their expectations they spend less, and the economy may settle at a lower level of economic activity (although, Mandeville suggests, at a more peaceful one). Mandeville also suggests ways of shocking people out of their lethargy, (which are discussed later in the section on the nature of government expenditures). The similarity with Keynes's views on "animal spirits," as well as with the formal model in chapter 1 is again obvious:

> Let us examine then what things are requisite to aggrandize and enrich a Nation. The first desirable Blessings for any Society of Men are a fertile Soil and a happy Climate, a mild Government, and more Land than People. These Things will render Man easy, loving, honest and sincere. But they shall have no Arts or Sciences, or be quiet longer than their Neighbours will let them; they must be poor, ignorant, and almost wholly destitute of what we call the Comforts of Life, and all the Cardinal Virtues together won't so much as procure a tolerable Coat or a Porridge-Pot among them: For in this State of Slothful Ease and stupid Innocence, as you need not fear great Vices, so you must not expect any considerable Virtues. Man never exerts himself but when he is rous'd by his Desires: while they lie dormant, and there is nothing to raise them, / his Excellence and Abilities will be for ever undiscover'd . . . [pp. 183–84]

> Would you render a Society of Men strong and powerful, you must touch their Passions. Divide the Land, tho' there be never so much to spare, and their Possessions will make them Covetous: Rouse them, tho' but in Jest, from their Idleness with Praises, and Pride will set them to work in earnest: teach them trades and handicrafts, and you'll bring envy and emulation among them . . . [p. 184][31]

(The formal proofs and mathematical translations for both Keynes's and Mandeville's views can be found in appendix 1.2.) What the quotations in this section make clear is that there are similarities among Keynes's, Mandeville's, and my views of human nature, on the "paradox of savings" in particular. These do not seem at all related, however, to the standard IS-LM models or to the departure points in standard micromodels; they are based on a different view of human nature. It is this view of human nature that can shed light on circumstances in which the intervention of government, through increased expenditures, may have a positive effect on so-called economic activity.

Measured Economic Activity and Government Expenditures

Keynes's view that an increase in government expenditures and in deficits can have a positive effect on *total* measured economic activity comes through clearly in the *General Theory*, where he implies that in some circumstances the increase does not necessarily crowd out private investment. In what circumstances may this be true? One set of circumstances has been clarified in the first section; a second set will be clarified here.

One does not have to go very far or do any sophisticated econometric research in order to find some obvious examples of no crowding out: it is well documented that preparation for wars and expenditures on arms succeed in ending depressions, diminishing unemployment rates while increasing deficits. Since the increased government expenditures spent in these cases certainly fit the definition of "public works," there is little doubt that there can be a positive correlation between government expenditures and people exerting themselves more, for total measured economic activity to increase (let us neglect for the moment discussions on long-run effects). The question is therefore simple: is it only the idea of war that can "touch people's passions" and lead to such outcomes, or are there also other cases in which a positive correlation between increased government expenditures and total measured economic activity can occur? The general answer given below will be that when the public's perception is that some ventures that *only* governments can provide are not provided, their provision by the government may *increase* private investment and consumption by restoring confidence. The key word in this phrase is "perception."[32]

Before reaching this conclusion, it helps first to recall the vocabulary used in the literature of government finance in the context of deficits. The literature suggests that a distinction should be made between "current account" and "capital account" when looking at budgets and deficits. A pure current-account expenditure is for a service of perfectly perishable goods that gives rise to no government-owned asset (and thus produces no future benefits). A pure capital-account expenditure is a purchase of a durable asset that gives the government a future stream of returns the value of which is greater or equal to the present cost of acquiring the asset. Government debt issued to finance a pure capital-account deficit does not lead to expectations of future increases in taxation (and even classical economic theory justifies such deficits), while debt issued for financing a pure current-account deficit does. For current-account deficits do not generate future benefits—it appears that these are the deficits scorned by some economists. However, so far as capital-account deficits are concerned, the government is like a firm, borrowing to finance projects that over time are expected to be profitable. The problem with this vocabulary and these arguments is, as illustrated below, that they do not provide a practical framework for discussing either deficits or crowding out.

Consider first the following two facts: (1) of the $350 billion spent by the U.S. government in 1984, $250 billion are expected to be spent on arms; (2) recent studies by the U.S. Science Council, NASA, and the Pentagon concluded that the main reward of the space program is the "psychic reward" of knowing the country is ahead of the Russians.[33] If so, how should one count the $250 billion—how much toward the current and how much toward the capital account? And how can one evaluate the stream of benefits with any precision in terms of the current and future "psychic returns" of knowing and expecting to be ahead of the Russians? Nobody can provide precise quantitative answers to these questions: the calculations depend on people's perception of the threat. This evidence suggests, therefore, these insights into the problem of deficits: (a) that distinctions between "current" and "capital" accounts may not be useful in shedding light on the issue. Thus in the rest of the analysis this distinction will not be made; (b) the distinction that will be made is whether the perception is that the government's expenditures are spent the way the public expects, or that the government mismanages its revenues. This is indeed the distinction made in a recent editorial in the *Wall Street Journal*, which gives the following interpretation of the 1984 budget:

> The emphasis in this budget, as in others in this administration, is on national security. As the president said in his State of the Union address, national security is the most important federal responsibility and he is giving it his highest priority. Indeed, the entire budget is noteworthy for its emphasis on what can be broadly described as investment. Scientific research (which gets a generous 15% boost), including even that manned space station, is an investment in the future. So is education and training, if properly managed. (One can also add medical care, that expenditures on it will reach $100 billion). So are efforts to save friendly countries from communist takeovers. And so is the defense budget, again, if properly managed. One reason for the strong dollar has been the flow of foreign investment in the U.S. and one reason for this investment is the fact that this country promises to remain relatively secure for the long-term. Its willingness to spend heavily on defense isn't the only reason, but it is one reason. [2 February 1984]

In other words, when perceptions of threats change, increased government expenditure on defense may promote, rather than crowd out private investments—some investments had been crowded out by the changed perception of threat *before* the government intervened. What institution, if not the government through a decision by a political entrepreneur, can gamble on a new defense strategy, hoping to restore international stability? In the absence of such a decision, the economic situation is worsened.

A similar analysis can be made of another major component in the budget: welfare payments, social-security benefits, and other expenditures broadly identified as "safety nets." As indicated in chapter 2, Bismarck was the first leader in modern times to implement such policies, the rationale being the fear of social instability. These policies were supported by the relatively rich, too (not only by the future recipients of the transfer payments), whose perception was that such insurances diminish threats of social instability and will encourage private investment. The conclusion with respect to these items in the budget is therefore the same as that with respect to military expenditures: only governments can provide such insurance, and the increased government expenditures in *these circumstances*, even if temporarily financed by increased deficits, may not crowd out private investment but encourage it (leading to expectations of balanced budgets, if managed properly). Both examples illustrate that

1. discussions as to estimating the future stream of returns from these types of government expenditures or determining whether or not some expenditures fit into "capital" accounts have no practical implications whatsoever.

2. when one discusses increased deficits the only thing that matters is people's *perceptions* of the ways governments spend the increased revenues.

Before examining the policy implications of these views, it may be useful to remind the reader again of statements Keynes makes at the end of the chapter on the marginal propensity to consume and the multiplier and to point out the view of human behavior that is lurking behind them:

> Pyramid-building, earthquakes, even wars may serve to increase wealth, if the education of our statesmen on the principles of the classical economics stands in the way of anything better. [p. 129]

> Ancient Egypt was doubly fortunate, and doubtless owed to this its fabled wealth, in that it possessed *two* activities . . . pyramid building as well as the search for the precious metals . . . The Middle Ages built cathedrals and sang dirges . . . [p. 131; italics in original]

> Just as wars have been the only form of large-scale loan expenditure which statesmen have thought justifiable, so gold-mining is the only pretext for digging holes in the ground which has recommended itself to bankers as sound finance. [p. 130]

These ideas seem quite strange, at first sight, and cannot be easily reconciled with either the neoclassical or the IS-LM approaches. But they can be linked in a straightforward way to the model and the arguments presented above. Let us see how.

While Keynes's "recommendation" of "making" earthquakes as a means to increase wealth certainly lacks common sense, he may have been right in the perception that after earthquakes a society's measured wealth may increase. For, not only is much of the society's physical wealth possibly destroyed after an earthquake, but, if many people die, its social institutions are likely to be destroyed, too, and its customs weakened. All these events and consequences imply both a worsened economic condition and more frequent fluctuations in people's position in the distribution of wealth. Both encourage people to bet more frequently on new ideas, the lucky ones leading to increases in measured wealth (for the elaboration of the mechanism see Brenner 1983b and 1984, and, reaching the same conclusion from alternative viewpoints, Olson 1982 and Ayres 1944).

Keynes's other examples are more subtle: after all, decisions to build pyramids and cathedrals or spend on arms are a matter not just of "public works" or "government expenditures" but of beliefs. People must *believe* that such expenditures contribute to their wealth. If people did not believe in an afterlife and were not religious, expenditures on pyramids and cathedrals would be perceived as diminishing wealth, rather than increasing it (and leaders advocating such expenditures, when the population was against them, sooner or later would fall). Similarly: major government expenditures on arms and on preparation for war cannot be considered "just" expenditures. If people did not feel threatened by foreign powers, such expenditures would also be viewed as diminishing "wealth," "morale" would be lacking, and, again, leaders pursuing a strategy of increased expenditures on arms when most of the population was against it, would probably fail. In other words, what Keynes's examples suggest is that whether or not increased government expenditures increase or diminish wealth depends on people's *perception* of those expenditures—a perception which is absent from discussions in today's "macro" models, but which plays a central role in my model.[34]

This view of the problem of deficits and of government expenditures has some clear-cut policy implications. Since one cannot tell with precision exactly what idea to pursue, or just how much should be spent on one form of expenditure or another, political leaders play an obvious and central role by trying to guess the wishes of the majority and then finding practical methods to implement them. If their guess is accurate, and if they succeed in changing the perception of risk (associated with either international or national stability), *no* crowding out will occur. But if their guess is inaccurate, and the majority's perception is that the expenditures have been wasted, crowding out will occur. In other words, political leaders may float ideas, see the consequences, and react. Some ideas will turn out to be successful, and others not; on some items there will be under-, while on others over-shooting. Essentially, this is a recommendation for a "stop and go" policy: suggest ideas, implement them, reexamine their consequences, and revise them, if necessary.[35]

Some recent opinions on government expenditures and deficits in Canada illustrate some of the implications of this view. A 1983 opinion poll carried out by Decima Research Inc. for the Ontario Economic Council reveals that 21% of the public thought that reducing the deficit would help economic performance a great deal, while 46% thought that it would help somewhat.[36] When asked what method would achieve that end, 65% suggested reducing services, and only 20% suggested increasing taxes. When asked which expenditures should be diminished drastically the answer was clear: the public wants to maintain educational and social-welfare programs (86% were against reducing access to health care, 74% against canceling family-allowance payments, 75% against cutting subsidies for education, 65% against cutting spending on job-creation programs, 55% against reducing passenger rail and ferry services). But 83% were for reducing the number of people working for the government, 55% for reducing postal service to three days a week, 50% for eliminating all financial support for Canadian Broadcasting Corporation and 50% for drastically reducing the number of people who would be eligible for unemployment-insurance benefits. These opinions suggest that the key objection to deficits stems from disagreement on *types* of government expenditures and on the way such expenditures are managed, rather than on government expenditures per se. The harm attributed to deficits in the standard economic literature, which looks at aggregates—whether high interest rates, or total government expenditures—is not the real problem. The harm comes *not* from the "deficits" but from the majority's *perception* of what happens to their wealth as a result of the government spending that the deficits help to finance.

Further Policy Implications

The discussion on Keynes's various views has identified some circumstances in which an increase in government expenditures can have a positive effect on measured economic activity. These circumstances can be divided into two broad categories:

1. When governments spend on projects linked with either defense or social stability, even if they are financed by deficits, they may not crowd out private investment. On the contrary, they may encourage it. Such policies, by definition, are nationalistic in character.

2. In the past, when decision makers within institutions like central banks made mistakes that inadvertently led to unexpected decreases in the price level and to bankruptcies, governments were able to correct them.[37] In such circumstances political entrepreneurship may also substitute for private efforts that have been eliminated by the mistaken policies.

The discussion in this study suggests that a sharp distinction must be made between these two cases: one has to do with the correction of an error, while the other is more fundamental and has to do with human behavior. Thus it is important to emphasize two things: first, that one should *not* confuse the mechanism underlying human behavior with the noises (i.e., errors) to which this mechanism sometimes leads; and second, that one should not examine the issue of government expenditures from just one perspective.

The discussion also suggests that today's problems may not be linked with "deficits" at all but may result from the majority's *perception* of two features. One involves the items on which the government spends its money—regardless of what the method of financing is, borrowing or taxing. Spending on items that the majority is against will produce the same harmful effects no matter how it is financed (though the channels for taxing and for borrowing are somewhat different).[38] When the public is against the expenditures, their perception is that they are wasted, a perception that leads to expectations of *diminished wealth* (no matter how some statisticians or economists measure the effect of the expenditures on the GNP): *It is this perception of wealth that matters, and not how the expenditures were financed.*

The second feature involves the response of the private sector to some new policies provided by the government through deficit financing. Even if the majority is in favor of these expenditures, the question is: Are they as optimistic about their positive effect on private investment as the government's? If they are not, they may be uncertain as to how the government will eventually solve the problem: by raising taxes (and if so for whom?), by inflationary financing, by continuous borrowing, or by diminishing expenditures. Each of these policies leads to different expectations as to its effect on the distribution of wealth, and thus as to future demands and supplies.

Last, but not least, the discussion in this section suggests that the current framework of aggregate, mechanistic models cannot shed light on any, so-called, macroeconomic subject. But an approach that looks at the problem by examining some fundamental characteristics of human behavior may do so.[39] The next section examines how the approach pursued above can shed light on additional phenomena associated today with macroeconomics.

3. The Phillips Curve, Rational Expectations, and Changes in the Distribution of Wealth

The short-run Phillips curve is one of the cornerstones of the now-popular "macroeconomic" theories. Its main postulate is that only unexpected changes in the inflation rate will lead to deviations from the "natural rate of unemployment"; when the inflation rate unexpectedly increases, the unemployment rate is assumed to diminish, while if it unexpectedly diminishes, the unemployment rate is assumed to increase. The Phillips curve has been an empirical finding for many, although by no means all, inflationary periods (certainly not for the hyperinflations), and various theories have tried to explain it.

Keynes's explanations for such relationships, presented in the previous sections, referred to very restrictive circumstances. Fisher's "debt deflation" theory predicts a similar relationship in the same restrictive circumstances that Keynes had examined in parts of the *General Theory*.[40] More recent models predict a Phillips curve by reliance on signaling models (Lucas 1973, for example), or by variations on Keynes's views that rely on the existence of nominal wage contracts (Fischer 1977, for example).[41] According to these recent models, unexpected monetary expansion (leading to unexpected inflation) always leads to increases in output, a prediction that has been frequently contradicted.[42]

According to the model and the arguments presented here one should *not* put all inflationary periods and all monetary expansions in one basket and expect that the effects of unexpected inflation on output and employment will be the same (*even* if the monetary expansions have the same effect on the price level). Let us reemphasize why.

First notice that in the model presented in chapter 1, one must distinguish between regular, temporary, short-run fluctuations in supply and demand and those that are unexpected (i.e., types of fluctuations that have not occurred in the past). The first type of fluctuations are already incorporated in people's expectations: in their perception of wealth and of the distribution of wealth and in their view of what a customary rate of unemployment might be (since probabilities have been assigned to finding certain types of jobs within certain time limits). The effects of fluctuations that the model can examine are those that are unexpected and significantly change the distribution of wealth: as described in the previous chapters and sections, examples of such fluctuations abound. Thus the "inflation-unemployment" trade-offs that can be examined within my model refer to cases in which such unexpected, significant changes in the distribution of wealth occurred and in which as a result the unemployment rate became significantly different from the customary one.

Let us examine the various interpretations that can be given to an inflation-unemployment trade-off in my model. Consider a situation in which expectations are frustrated and the distribution of wealth changes. In such a case the incentives to gamble on entrepreneurial acts increase, and productivity and output also increase. Simultaneously, the incentives to pursue inflationary policies also increase: either because the value of the government's internal or external debt has increased or because some groups, who have become worse off, put increased political pressures on the government. Thus the data may show an increased output, diminished unemployment, and increased inflation—and these are the characteristics of the Phillips curve. Yet the curve, according to these arguments, reflects a correlation and not a causal relationship.

These arguments provide a straightforward explanation for the persistence of the effects on output of some unexpected changes, an explanation that "rational expectations" models have difficulty in providing. People continue to make greater efforts and engage more frequently in entrepreneurial acts until their position in the distribution of wealth is restored to its expected level. How long this process takes, we do not know: it depends on the number of "lucky hits" made, on how much the realized wealth has deviated from the expected one, on the changes in the power of organized groups, and on the changes in fiscal policy. The roles of the first three factors have been explained; the fourth, the government's fiscal policy, also plays a role, since if along with frustrated expectations the government is simultaneously perceived as pursuing more redistributive policies, the incentives to engage in entrepreneurial acts diminish, and wealth will be restored to its expected level more slowly (such adaptation processes may be defined as "business cycles").

Yet, as shown in the previous section, governments and central banks may gamble on inflation not only when they want to impose a tax or to redistribute wealth, but also when, in a situation worsening without anyone knowing why, they bet on Keynesian ideas and believe that inflationary policies will provide a remedy. Whenever the inflation rate deviates from the expected one, the expansionary monetary policy has a *harmful* effect on wealth: whether the inflation rate unexpectedly increases or decreases, output and employment diminish. Let us see why.

As discussed in the previous section, the existence of nominally denominated long-term contracts characterizes a monetary economy. However, whenever the realized inflation rate deviates from the expected one, there is a dead-weight loss, according to Keynes because of the way the labor market behaves, while according to Fisher, because of a more complex argument, which is related to the theory of the firm presented in chapter 2.

Suppose that an inflation rate was expected in the economy, that the rate of interest was adjusted to this rate, and that firms were net borrowers and consumers net savers. If the inflation rate unexpectedly diminishes, the real value of the firms' debts increases unexpectedly, profits diminish, and the number of bankruptcies increases. Two types of costs are imposed on the economy. Bankruptcy costs redistribute income from firms to lawyers and accountants. The second cost arises from the misallocation of entrepreneurs, and it is impossible to evaluate since it is due to forgone opportunities. The entrepreneurs lose their jobs not because they have been bad managers, but because the value of their firm's debts has increased due to the unexpectedly changed government policy. Here is Fisher's (1933a) comment on the process:

> [A bond holder] and the other creditors are compelled to go into business by taking over the business that failed . . . But they were never cut out for such a work . . . That is one of the many damages that affect society as a whole; [and] at the end of a deflation period the wrong people are in charge and recovery is just as much delayed, and less successful when it comes . . . [p. 64]

The addition made here to Fisher's views[43] (which are presented in detail along with his "debt-deflation" theory), is the precision to his discussion of the role of "entrepreneurs" in the theory of the firm.

What happens when the inflation rate suddenly increases? The value of the firms' debts decreases, and the accounting profits increase. Since profits provide one signal as to management's abilities, and stockholders cannot easily identify why profits increase, bad managers are lucky—they will be changed more slowly than if the price level were constant. However, it is impossible to evaluate the costs due to this misallocation of entrepreneurs (because of the slower turnover), since, again, the costs refer to forgone opportunities.

In conclusion, according to the model presented in this book, in some circumstances monetary expansions may be correlated with increases in output and employment (when their purpose is to impose a tax and redistribute wealth), while in other circumstances they may not be positively correlated (when people believe in mistaken theories). It is therefore a mistake to put all inflationary episodes in the same basket (even if their effect on the price level is similar).

4. On Inflationary Policies

The views of Keynes and Fisher on inflation-unemployment trade-offs have been discussed. More recent explanations for these trade-offs are based on the assumption that people make greater errors when the unexpected inflation is higher, and that nominal contracts and expectations are adjusted during a lag. The point of departure for these more recent models is that the "natural rate of unemployment" (NRU) is independent of monetary policy. The arguments presented in the previous sections suggest that this point of departure is accurate only if the inflation is due to a misunderstanding of the inflationary process, but not if the inflation has been chosen as a method of taxation and redistribution of wealth.

If central banks and governments gamble on inflationary policies when expectations are frustrated, the unemployment rate is not independent of the government's monetary and fiscal policies, since this rate *depends* on the distribution of wealth. Thus, in the model presented in chapter 1, when unexpected changes occur, nobody can know what "the natural" (?) rate of unemployment is: the NRU is not well defined since the distribution of wealth was altered (in this sense the model is a precise "disequilibrium" model).[44]

"Rational expectations" (RE) models have assumed that the NRU is independent of monetary policy because they have never raised the question as to what incentives have been changed in the economy so that governments and central banks have suddenly become more likely to gamble on new policies, inflationary ones in particular. In fact, these models avoid discussion of the effects of changes in the distribution of wealth. The impression left by these models is that inflation has only a temporary effect on relative prices and that it begins when governments decide, without apparent reason, to raise their revenues, and not because there might have been a change in the direction toward which the economy was moving. Inflation is thus perceived to be a costly caprice carried out against the wishes of most voters. The best thing to do is to follow rules and eliminate the caprice. However, if inflation and changes in government policy are due not to caprice but to some unexpected events that lead to frustrated expectations, the policy implications are not so obvious.

If indeed the latter argument explains prolonged inflationary processes, one may argue that when political pressures come from one sector rather than another, the economy could be "better off" (in the sense of increased output, and *not* in the sense of utility) if governments and central banks were not susceptible to such pressure. Then more people would gamble on individual effort to restore their wealth, rather than on political effort which would just redistribute it. Thus the rate of increase in output could be greater and wealth would be more quickly restored to its expected level. But since governments are susceptible to political pressures (and it seems naïve to assume that they can work differently), the alternative to inflation would be either taxation or increased borrowing, just as if the internal debt had increased. In this case it is more difficult to infer that inflation is necessarily a worse strategy than imposing a tax (a "capital levy," according to Keynes's [1923] suggestion), since legislation of taxes and their implementation takes time, a resource that governments frequently lack.

Only if inflations were correlated with beliefs in mistaken theories would the conclusion that we would be better off without them hold—very obviously—true. But the fact that major inflationary episodes have been correlated since antiquity with difficult times suggests that they have been chosen for reasons other than belief in incorrect theories. Also, the fact that after wars governments have frequently gambled on inflation suggests that policy makers seem to perceive (perhaps wrongly) that the benefits of inflation in these circumstances outweigh its costs. One may ask what the alternative is for the inflationary gamble when time is too short for implementing a well-thought-out taxation scheme. Confiscation is certainly one alternative; the reason politicians in democratic countries avoid this option may be that they realize that such a policy could put an end to their careers. In contrast, the identity of the person who has gambled on the inflationary policy is less evident: one can always blame OPEC, the unions, or the previous administration.

In conclusion, both the traditional views on inflation and the one presented here imply that when inflation unexpectedly increases, output may increase as well, and that only unexpected changes "matter." But the views differ in their identification of the unexpected change: the standard approaches refer to the money supply, while the one presented here refers to the changed incentives when expectations are frustrated. There is one minor issue that this model cannot examine while the alternative approaches can: namely, the one related to the rate of increase in the price level. The model presented here only predicts the circumstances in which it is more likely that central banks will gamble on inflationary policies, and what might then happen to output and employment. The traditional approaches examine the question of what might happen to the price level *if* the inflationary policy is chosen, an issue on which there seems to be little, if any, disagreement. However, the next chapter shows how the two approaches can be combined to shed light on recent events in Canada.

The fact that one can explain and understand when pressures for inflationary policies are more likely to occur does not imply preference for the inflationary strategy. But since in this model inflation is at times viewed as one method of redistributing wealth, it is not obvious that in all circumstances inflation is the worst option (are revolutions better?!). In some unexpectedly altered circumstances (prolonged wars, increased government debt, an unexpectedly lowered price level due to mistaken policy, etc.) the majority may perceive inflation as the best policy option—"best" in the sense of restoring or maintaining stability. In other circumstances, when decision makers in governments bet on the mistaken ideas of some economists, the majority may perceive inflation as harmful and the inflationary policy may have several destabilizing consequences. Since in the latter case wealth is arbitrarily redistributed (arbitrarily from the majority's viewpoint), those who fell behind may either advocate restoring the customary redistribution (by imposing temporary surtaxes on the suddenly increased profits due to the lowered value of the debt of enterprises) or make revolutionary demands about the distribution of wealth. With such consequences inflationary policies cannot be perceived as good ones—they have totally unpredictable consequences, since nobody can predict the precise nature of the revolutionary demands that may capture people's imagination.

Last, but not least: there is no morality in the arguments presented in this chapter either. But then there is no morality (at least not from an outsider's viewpoint) in wars (which almost without exception have been followed by inflations) or in an initial distribution of wealth for whose restoration inflationary policies are at times pursued.

Appendix 5.1:
Unemployment: A Note

Today economic models explain "unemployment" in several ways: (a) one model defines the "natural rate of unemployment" as one that depends on minimum-wage laws, union restrictions, structural changes; (b) another model attributes part of the unemployment rate to the fact that searching for jobs takes time; (c) still another attributes part of this rate to the intertemporal redistribution of time. The common characteristic of these three models is that within them government policies cannot consistently affect the unemployment rate (see Phelps 1970). In contrast to these three types of models (associated with "monetarism"), there are others that rely on the Keynesian concept of "involuntary" unemployment, within which government policies can have a permanent effect on the unemployment rate.

The characteristic common to all the aforementioned models is that customs and the political "market" (i.e., the decisions of politicians) play no explicit role in them. In contrast, in my model these factors play the central role, and "monetarism" and "Keynesian theories" represent nothing more than two guesses that some economists have made to explain some facts without relating them explicitly to a society's political life.

Yet one may ask how the term "unemployment" can be defined within my model (as shown in chapter 3, many words that we are accustomed to—profits, entrepreneurs, productivity, etc.—must be redefined within it). The unemployment figures published today are defined from a bureaucrat's viewpoint: laws and regulations define them. Within my model the word "unemployment rate" can be defined as "the fraction of people unable to continue their customary employment (or 'way of life')." As shown below, the suggestion that these words be used to define the unemployment rate is not a mere play on words but has definite consequences (in contrast to the Russian style where the "unemployed" are called "hooligans," an implication that there *is* no unemployment in Russia; since the statistics on crime are not published, the redefinition not only eliminates "unemployment," but does not increase the "crime rate" either).

In traditional macroeconomic theories the central problem is how to explain why decreases in demand cause unemployment rather than immediate wage and price adjustments in labor and nonhuman resources (the "search models" have been developed to provide an answer). But answering this question seems easy when looking at the definition of "unemployed" within my model. For a variety of reasons customary ways of life have been abandoned; people now marry later and have fewer children, and the divorce rate is higher than in the past. Women, for example, have abandoned their customary way of life (as housewives and mothers) and have decided to work outside their homes (did they have an alternative?). When such changes take place it is not so much "wage and price adjustments" that can explain the temporary increase in the unemployment rate. Rather, *the time* it takes until *entrepreneurs* appear who perceive the new opportunities (inventing and selling dishwashers, washing machines, etc.) which will, eventually, provide more employment outside the home may explain the increase. This same argument for the increased unemployment rate can be repeated for any occupation in which people have specialized for relatively long periods of time expecting stable circumstances, but suddenly face drastic changes (recent developments in the U.S. auto and steel industry come to mind, as well as the effects of the change in import prices).

Does the redefinition of words suggest an insight as to whether or not the government can provide a solution? As implied by the model and discussed in the text one should distinguish between two cases: Suppose that a customary demand has dropped. Politicians may now impose an implicit or an explicit tax to subsidize the affected sector in the expectation that perhaps the disturbance is temporary. Suppose that the amount of the subsidy is $200 million per year and there are 50 million people employed. On average a $4 tax per year does not change people's attitudes toward risks, in my model, which means that levels of effort will not be significantly changed. What happens in the long run is another matter (as discussed in the text, such redistributive policies have dangers), but in the short run the government can increase employment and maintain stability. In this sense the government can "create" employment (or, more correctly, can prevent employment from diminishing significantly).

Can the government (i.e., some politicians) create employment by other methods? As shown in this chapter, that depends on the specific historical circumstances. When customs are weakened employment will be provided by entrepreneurs who gamble on new ideas. Does the government (i.e., the political entrepreneurs) have the advantage of carrying out such gambles? Sometimes not: for not only do politicians not have a relative advantage over others in perceiving new business opportunities, but they may do even worse than other entrepreneurs since it is with the taxpayers' money that they gamble. But sometimes, when perceptions of risks associated with either national or international stability have changed and have crowded out private investment, governments can implement new strategies that restore stability.[45]

6   What Insurance Can Indexation Provide?
The Canadian Experiment

With every major outburst of inflation discussions on indexation became popular. While the idea is theoretically appealing, its implementation has always been problematic since, as illustrated in this chapter, either the index numbers chosen have led to unexpected redistributions of wealth, or inflation may have been perceived as a political necessity (due to a suddenly increased government debt). I shall try to show, first, why the idea of indexation is appealing within some theoretical frameworks, and second, by surveying Canadian and other experiments with it, the practical difficulties that have *always* been linked to the issue of the redistribution of wealth.

1. Historical Background: Money and Indexation[1]

The concept of indexation was developed along with the concept of index numbers. But, while the latter concept led to the belief that "economic theory is not only quantitative but also numerical"[2] and had a significant impact on both economists and the public, the former remains controversial to this day. With a review of the discussion on the subject, it becomes evident that the concept of indexation is in many respects consistent with the monetarist view of the economy.

The central problem of monetary theory before Keynes's *General Theory* was not unemployment but the value of money. At that time research interests in this field took off in two directions: building index numbers and elaborating the quantity theory of money. The idea of indexation was an outgrowth of these studies.

Since price-index numbers date only from 1860, until the nineteenth century discussions of ways to link wages and debts to some prices could not have been of practical importance. Yet the idea of defining general purchasing power was known earlier, as is evident from Bishop Fleetwood's Chronicum Preciosum (1706):

> . . . 'tis evident, that if £5 in H.VI. days would purchase 5 Quarters of Wheat, 4 Hogsheads of Beer and 6 Yards of Cloth, he who then had £5 in his pocket was full as rich a man as he who has now £20 if with that £20 he can purchase no more Wheat, Beer, or Cloth than the other. [quoted in Keynes 1930, p. 49]

Fleetwood suggested that the value of money could be measured by a set of commodities and that this practice might be useful in keeping wages constant in real terms. However, the idea of index numbers and indexation developed after the Napoleonic wars, when England faced volatile price changes: an inflation of 60% between 1792 and 1814, followed by a deflation of 40% until 1822. It is Joseph Lowe (1822) who improved the technique of measuring index numbers and recommended their use:

> It would be in itself merely a table of reference, and all contracts, whether relative to loans, leases, or bequests, might, at the will of the parties, be made payable, either according to the proposed standard, or in money of undefined value . . . And what would be the practical application of this knowledge? The correction of a long list of anomalies in regard to rent, salaries, wages, etc., arising out of unforeseen fluctuations of our currency. [quoted in Fisher 1934b, p. 25]

Here indexing is viewed as a tool for eliminating the risk of unanticipated changes in the value of money.

According to Schumpeter,[3] it was G. P. Scrope who introduced the subject in a general analysis in his *Principles of Political Economy* (1833), his emphasis being not on protection against inflation but on preventing the deflation-induced rises in the value of debts as they "may covertly despoil of value . . . the classes whose industry or property is embarked in productive operations, in the benefit of those who own monied obligations" (quoted in Fisher 1934b, p. 28).

These were not the only opinions at that time. Others saw the problem of inflation within the broader context suggested in the previous chapter, where it is recognized that indexation cannot be recommended in all circumstances:[4]

> . . . the variations in the purchasing power of money brought up the question of "justice" as between creditors and debtors (or else, as far as the public debt was concerned, taxpayers). As always, "justice" was what benefited the interest with which each writer sympathized . . . "Squire" Western made the point that there are situations in which higher prices being the only alternative to widespread bankruptcy, a falling value of money might be deemed to be in the interest of the creditors. [Schumpeter 1954, p. 713]

Later Jevons (1876) advocated the indexation scheme—then called "the tabular standard"—of Lowe and Scrope and even suggested imposing it by law since "it would add a wholly new degree of stability to social relations, securing the fixed incomes of individuals and public institutions from the depreciation which they have often suffered . . . Periodical collapses of credit would no doubt recur from time to time, but the intensity of crises would be mitigated, because as prices fell the liabilities of debtors would decrease approximately in the same ratio" (p. 333). Marshall, too, in his "Remedies for Fluctuations" (1887), advocates the tabular standard. He returns to the ideas of Lowe, Scrope, and Jevons, recommending the introduction of the tabular standard as a means of preventing credit cycles.

Indexation was also advocated by Edgeworth and Jevons as a consequence of their view of the "Indefinite Standard" which they defined as the measure of the changes in the monetary unit, "a simple unweighted average of the ratios formed by dividing the price of each commodity at the later period by the price of the same commodity at the earlier period" (quoted in Fisher 1934b, p. 39). The practical problem was to find such an index. They realized that that was not easy since not all money prices change at the same rate. Yet they thought that

> . . . the fluctuations in the prices of individual things are subject to two distinct sets of influences—one set due to "changes on the side of money" which . . . affect *all* prices *equally* in direction and in degree, the other set due to "changes on the side of things" which affect prices relatively to one another. Now as regards the second set, changes in the prices of things relatively to one another can involve no absolute change in the value of money itself. . . . By a change in the value of money in itself or in the price level as a whole, they mean the amount of the uniform residual movement due to "changes on the side of money" after we have "averaged out" the chaotic but compensatory movements in individual prices due to their movements relatively to one another and to the price level. [Keynes 1930, pp. 72–73; italics in the original]

This "averaging out" is the basis for the "Indefinite Standard" that brought Jevons and Edgeworth to the idea of indexation as a way of identifying "purely monetary" changes and providing an "objective" measure for the value of money.

This last point was elaborated by Irving Fisher (1911) in *The Purchasing Power of Money*. In spite of the fact that he considers the problem of different price indexes for different groups of commodities, he concludes that, though theoretically such distinctions may be important, practically they are not. Therefore, "although different persons and different classes may establish different standards for special contracts . . . a single series of index numbers including articles used and purchased by all classes, and including also services would probably be found advisable" (p. 217), and "on the whole the best index number for the purpose of a standard of deferred payments in business is the same index number which we found the best to indicate the changes in prices of all business done; in other words it is P on the side of the equation of exchange" (p. 225).

These quotations show the relationship between indexation and the quantity theory of money: the index suggested as a basis for the deferred payments being the P of the equation of exchange. But according to Fisher the proportionality between the quantity of money, M, and the price level, P, holds only in states of equilibrium and not during transitional periods during which consumption and investment patterns are altered *because* of expectations of increased money supplies in the future (I shall return to this point below when examining the Canadian evidence). Thus during such periods no distinction can be made between the effects of "purely monetary" and "real" changes on any price index, and the indexation scheme may not be easy to implement. Why doesn't Fisher mention this problem when he discusses indexation? The reason may be that he was interested in deriving some simple policies from his theory.

To summarize: the principal objective of indexation was and still is perceived to be (see appendix 6.1) to prevent the distributional effects of unanticipated changes in the economy and capture those changes that were "purely monetary." Yet the advocates of indexation avoid discussing two questions. First, how accurately can price indexes capture such changes? and second, should everybody be protected against the effects of monetary changes, or is their very purpose to redistribute wealth? As the evidence discussed next suggests, the problems with indexation were due to the fact that price indexes could not accurately capture "purely monetary changes" and in some circumstances the monetary changes were linked to sudden increases in government debt. The rest of this chapter discusses the evidence by concentrating on Canada's experience (the experiences of other countries are mentioned only briefly since they have been discussed at length elsewhere), and points out the intended and unintended consequences of the indexation of wage contracts and of transfer payments, consequences that have been linked to the redistribution of wealth.

2. Indexation: The Canadian Experiment[5]

Since the early 1970s Canadians have sought to protect the purchasing power of their incomes against erosion by accelerating inflation. During this time, linkage of increases in incomes to overall increases in prices in the economy—called "indexation"—has become an increasingly widespread practice. Transfer payments (such as family allowances, old-age-security payments, child-tax credits, wage contracts and personal-income taxes) have all been adjusted for inflation. The size of the adjustment has been determined by the consumer price index (CPI), which is considered the most appropriate measure of inflation.

Indexing of incomes in this way is regarded to have both benefits and costs.[6] The practice is perceived as beneficial because it is expected to maintain the purchasing power of incomes and reduce the costly and laborious transactions involved in frequently renegotiating contracts, adjusting personal-income-tax exemptions and tax brackets, and revising regulations. Indexation has costs as well, however: when incomes are fully tied to increases in the overall rate of inflation, temporary price shocks can arbitrarily change the distribution of wealth and, as explained in the previous chapter, lead to unexpected disturbances, increasing political pressures to pursue inflationary policies, in particular.

In the first part of this section indexation is distinguished from alternative methods of inflation protection. The use of the practice in private-sector labor markets and in the public sector is then examined. The third part of this section shows that the use of inaccurate price indexes causes an arbitrary redistribution of wealth and may lead to increased unemployment and increased government deficits. In the fourth part some alternatives to remedy the unintended effects of indexation on the distribution of wealth are discussed.

The Practice of Indexation

Indexation is a method of protecting incomes against inflation by linking them to increases in the overall price level of the economy. An alternative method of protection is to guess what the future rate of inflation will be and renegotiate contracts more frequently. This alternative assumes, however, that the power to renegotiate successfully exists and that people are able to guess future inflation correctly. If this is the case, a $10 payment when there is no inflation will become $11 when 10% inflation is anticipated. If the guess is incorrect, however, one of two things may happen. Either purchasing power will diminish or the contract must be renegotiated—a process that often occurs with considerable lag. With indexation the inflation protection is achieved automatically when inflation rates vary significantly. The $11 payment results from a previously negotiated contractual agreement that the payment will increase if inflation occurs. As a result, indexed contracts can be entered into for a longer duration than nonindexed ones—the evidence discussed next shows that this has indeed happened.[7]

Wage Indexation in the Private Sector. Wage indexation has been practiced with varying frequency since World War II. The practice is widespread at the present time, and it is commonly perceived to *contribute* to inflationary pressures by increasing labor costs faster than productivity gains. Yet the findings presented in this section show that recent indexation clauses have provided only a partial protection against changes in the CPI, and gains due to indexation have been less than the growth in productivity. Thus the conclusion reached is that wage indexation per se has not perpetuated inflationary pressures (although the overall increase in labor costs resulting from negotiated settlements that exceeded productivity growth, might have done so, as argued in the previous chapter).

Wage indexation became increasingly widespread during the periods of high and variable inflation in the early 1950s and again in the years since the early 1970s: tables 6.1–6.3 show the variability in the CPI and the variations in the extent of indexing practices during this period. As expected, while indexation clauses existed in 21% of all collective-bargaining agreements in 1951, by 1967 this proportion had dropped to 2%. However, while in 1969 only 9% of wage contracts contained indexation clauses, by 1975, 54% contained them.

Other economies have had similar experiences. There was widespread indexation of labor contracts in Brazil during the 1963–73 period when the inflation rate varied between 16% and 85%; in Israel during the 1948–82 period when the inflation rate fluctuated between 5% and 130%. In Belgium, Denmark, Finland, and the Netherlands the experience was similar during periods of high and variable inflation during the 1950s and 1960s.[8]

Why has wage indexing become so popular? For two reasons; as mentioned earlier, one reason is that it reduces the requirement of renegotiating contracts when inflation is higher than anticipated at the time of the original settlement: indeed table 6.4 illustrates that the length of contracts with cost-of-living adjustments (COLAs) has consistently exceeded those without COLAs:[9] in 1976, the average duration was approximately 35 months for the former and only 16 months for the latter.

A second reason is that indexed contacts are thought to exceed the value of nonindexed ones. On empirical examination, however, this perception turns out to be erroneous. As can be seen in table 6.5, while there were slight differences between indexed and non-indexed wages in the annual rates of change between 1970 and 1977, the averages for the period were almost identical: 11.8%.[10] Since the CPI rose during the same period by an average of 7% per year, real wages, indexed or not, grew by an average of 4.8%. This information is inferred from table 6.5. This increase in the purchasing power of wages indicates that indexing was only one component that contributed to wage increases. If the COLA had been the sole factor determining the size of the contract and had provided 100% protection against rises in the CPI, the purchasing power of wages would have been unchanged instead of increasing.

However, COLAs provided less than full protection for several reasons. First, in 36% of the agreements with COLA clauses, the clauses were "capped," that is, limited to a certain maximum amount. This arrangement meant that if the realized change in the CPI was 12%, but it was agreed that a maximum compensation of 10% would be awarded, then only a 10% adjustment was automatically paid. Second, in 27% of the agreements, the COLA clauses were "triggered" only after some minimum amount of inflation. This meant that for a 5% rise in the CPI no compensation was given, but if the rise was 10%, then a 5% automatic compensation was given. Third, in 60% of the agreements the COLA clauses did *not* cover the entire contract period. The fact that the percentages add up to more than 100% means that in some contracts *more* than one constraint was imposed on the COLA formula. These data are based on 1405 major collective agreements signed between 1968 and 1975. David Wilton (1980) has found that the degree of protection afforded by all the COLA clauses against changes in the CPI was only 47%.[11] This finding means that the COLA clauses guaranteed that, on average, for a 10% rise in the CPI wages would rise by only 4.7% and implied a 5.3% erosion in their purchasing power.

But productivity (as measured by output per man-hour) increased at an average annual rate of 4.4% for the years 1966–71 and 1.8% during the period 1971–75.[12] These increases in productivity combined with the 5.3% decrease in the purchasing power of wages due to the less than full indexing meant that wage increases *due* to indexation did not exceed productivity gains.

In summary, the fact that the wage increases were of the same size whether wages were indexed or not and that the productivity gains averaged 1.8% and the indexation protection was 47% suggest that wage indexation per se was not inflationary. This conclusion does not imply that the total increase in negotiated wages did not contribute to such pressures. It could have done so: suppose that at the beginning of a year wage increases were negotiated on the basis of expectations that there would be constant inflation. Suppose, then, that import prices suddenly rose (as oil prices did unexpectedly in 1973–74). In such a situation there are two possibilities: either wages have to be renegotiated and lowered to reduce costs, or costs will rise faster than prices, and unemployment will increase. Higher unemployment may result in increased pressure on the government to pursue a more expansionary monetary policy or to subsidize some sectors. Once governments *respond* to such pressures, additional sectors may try their luck, too, and, if they are successful, a process popularly labeled "wage catch-up" or "spillovers" among sectors will result. This is a process that can occur whether or not there is indexation, and, as argued in the previous chapter, it can lead to inflation.[13]

Indexation in the Public Sector

The policy issues related to indexation in the public sector are quite different from those in the private sector for several reasons. First, public-sector wages have had less automatic inflation protection than wages in the private sector. Second, legislation has been passed to index tax brackets and transfer payments *fully* to the CPI.

Wages in the Public Sector. As can be seen in table 6.6, indexation of wages in the public sector has provided less protection against changes in the CPI than have those in the private sector (with the exception of Quebec in 1972). In 1973, for example, COLAs in the public sector provided automatic increases that equaled only 25% of the increase in the CPI, in contrast to 38% in the private sector. Thus, if one were to find that wages in the public sector rose by as much as, or more than, those in the private sector, one would not be able to attribute the increases to "indexation."

Analysis of the reasons for larger wage increases in the public than in the private sector are not the subject of this chapter. It is useful to emphasize, however, that since COLAs provided only a 25% protection against changes in the CPI, the total wage increases have occurred *not* because of indexation, but either because the political power of public-sector employees has increased or because taxpayers have been ready to pay more for their services. Wage increases for either of these reasons should not be confused with wage increases due to indexation. It follows, therefore, that where public-sector wage increases are found to be "excessive," the remedies for adjusting these wages have to be found through factors *other* than public-sector wage indexation (by curtailing their power, for example).

Personal Income Taxes and Transfer Payments. Indexation seems beneficial in labor markets since it avoids the risks of guessing future inflation rates and of having to renegotiate labor contracts. Nonmarket contracts, like tax brackets, family allowances, and public-utility rates are also indexed—but for different reasons. These rates are usually determined in current dollars and are infrequently inflation-adjusted because of the length of time required to revise the necessary legislation or regulations.[14]

In the tax system distortions can occur during inflation, because incomes are automatically pushed to higher tax brackets even if the purchasing power of incomes before taxes has not changed. One result is that the distribution of wealth is arbitrarily changed. Indexation can prevent such arbitrary redistributions by adjusting tax brackets to increase at the same rate as the overall price index.[15]

Transfer recipients, such as those who receive family allowances and old-age-security payments, experience a reduction of the purchasing power of their benefits when an unexpected increase in inflation occurs. Indexing these transfers, therefore, diminishes or eliminates the lengthy legislative process and provides for automatic adjustments, as for wage contracts in the private sector. The Family Allowances Act, 1973, which came into effect on 1 January 1974, provided for Family Allowances and Special Allowances to be increased at the beginning of each year by the expected increase in the CPI (although in 1976 full indexation was suspended for a year as a result of the federal government's expenditure-restraint program).[16]

The Difficulties of Practicing Indexation

As the foregoing discussion has illustrated, indexation has been practiced in different ways in the public and private sectors. The rationale for indexing wages has differed from that of indexing transfers and income taxes, and the impact of these practices has varied. In all of these examples, the CPI has been the index used. The purpose of this section is to illustrate that many of the difficulties attributed to indexation relate directly to the arbitrary redistributive effects of the choice of the CPI as the index and to the decision to index transfers and tax brackets *fully*. Although the impact of these two decisions has not been assessed in Canada, other countries have estimated the magnitude of the undesired impact of fully indexing contracts to the CPI and have tried to correct for it. Lessons from their experiences are summarized at the end of this section.

The Choice of the CPI as the Index.[17] The use of the CPI for all indexation purposes in Canada may lead to several problems. The CPI

• overestimates the change in the cost of living;
• includes the effects of exchange-rate depreciation and of changes in sales taxes;
• measures not only changes in the prices of goods that are consumed every year but also changes in the prices of investment goods;
• treats changes in housing costs arbitrarily.

The CPI is a "fixed-weight" index, which means that it is calculated by means of a survey of prices of a large number of items typically purchased by consumers (a "basket"). The total cost of these items can then be compared with the cost of the same basket in a selected base period. Until the spring of 1982 these weights in Canada were based on a 1974 basket. (Now they have been updated to a basis of 1978 prices.) "The inflation rate" measures the percentage change in the cost of this basket, even though consumers change their consumption patterns to compensate for erosion in their purchasing power. For example, between 1967 and 1974 the inflation-rate calculation was based on a basket of goods in which food represented 24.8%, housing 31.4%, and recreation 6.9%. In 1974, when these weights were reestimated, it was determined that people were spending less on food, and more on housing and recreation. As a result the weights were changed to 21.5%, 34.1%, and 8.3%, respectively. Yet for seven years the changing consumption patterns were not reflected in the index used to measure eroding purchasing power. One of the central practical problems related to indexation is that the CPI overestimates the changes in the cost of living: a simple numerical example can be used to illustrate how. Suppose that the prices of gasoline and of heating oil unexpectedly increase by 25%, and that as a result people, on average, reduce their consumption of these items from 15% to 11%. Simultaneously, they increase their demand for warmer clothing in winter and lighter clothing in summer (because offices and houses are less heated in winter and less cooled in summer), for public transportation, fuel-efficient cars, and vacations closer to their homes. The calculation of the CPI does not take these substitutions into account. Instead, it is calculated as if people continued to spend 15% of their incomes on gasoline and heating oil. Thus, although this sudden, unexpected 25% increase in the price of gasoline and heating oil leads to a 3.75% increase in the CPI (15% multiplied by 25%), this number is meaningless in economic terms: it represents the increased cost of a basket of goods and services that people can no longer consume. Instead, they must adapt themselves to the increased scarcity (by the "gambling" process suggested in the previous chapters). Similarly, the CPI overestimates the change in the cost of living when the exchange rate depreciates or the prices of imported versus locally produced goods unexpectedly increase. Again, substitution from imported to locally produced goods is not taken into account.

Let us consider now the effect of changes in sales taxes on the CPI in

Canada. When the effects of these changes are included in the CPI, and these taxes are raised, the CPI increases. Yet this increase, too, is meaningless in economic terms: it represents a reallocation of resources in the economy to which people must adapt themselves.

Finally, the problems that arise because the CPI is supposed to measure things consumed every year but does not should be pointed out. The basket used for calculating the CPI includes durables like television sets and diamond rings which are not consumed in a year but represent, in part, investment and speculation. The fact that the calculation of the CPI confuses "consumption" and "investment" and that no price index can deal with speculations becomes especially troublesome in the CPI's treatment of housing during inflationary periods. The reason for the difficulty is that at any given time only a relatively small proportion of the population buys houses. With an increase in either mortgage rates or the price of housing the people who already own houses have not necessarily become worse off since the value of their asset may rise proportionately with the increased price of new housing. Thus the increase in the CPI due to the increased prices of new homes overestimates, too, the change in the cost-of-living.

The measurement of increases in housing and durable asset prices becomes even more complex in an inflationary period because it is during this time that incentives to speculate in such assets as inflation hedges increase. No index can appropriately deal with the shift in the perception of the value of houses and other assets used as hedges from "consumption" to "investment" goods. The CPI has this flaw, and it fails to take account of such a change in perception; as a result, it further overestimates the change in the cost of living. The information provided by the calculation of alternative forms of home ownership suggests how skeptical one must be when interpreting changes in the CPI during inflationary periods. It has been estimated that calculations of the annual inflation rate in the U.S. could range from 10.8% to 13.3% in 1979 and from 11.5% to 15.6% in 1980.[18] These wide variations show that even if wages rose by only 12% in 1979, when according to one calculation the CPI rose by 13.3%, this does not necessarily imply that the purchasing power of payments dropped by 1.3%. An alternative calculation of the CPI showed that the inflation rate may have only been 10.8%, in which case the rise of 12% in wages would imply that their purchasing power has increased by 1.2%. Thus, the fact that payments are less than fully protected against changes in the CPI does not necessarily imply a decrease in the standard of living.[19]

In conclusion, when there is inflation and people change their consumption patterns, the meaning of changes in price indexes becomes blurred, and more than one interpretation can be given to changes in the CPI. This conclusion indicates that the first difficulty with practicing indexation is the choice of an index. While only the difficulties in interpreting changes in the

CPI have been discussed here, it should be reemphasized that all other fixed-weight indexes would encounter similar problems.

Pitfalls of Full Indexing. The other main difficulty of indexing has to do with *full* 100% indexing in the public sector. In contrast to the partial indexing of wages that has been negotiated in the private sector, the federal government has fully indexed transfers and personal-income taxes. In Canada, between 1974 and 1982, personal-income-tax exemptions have increased to the full amount of increases in the CPI, and Family Allowances and the Child Tax Credit have also been fully indexed. Other countries with longer experience with indexation than Canada also began by fully indexing transfers to the CPI. Both the United States and Israel have linked social-security payments to the CPI, although neither country has indexed its tax brackets.[20]

Studies undertaken in the United States have shown that between 1972 and 1980 the CPI, to which social-security benefits were linked, increased by 12%, or 23% more than alternative indexes that were corrected for some of the measurement problems in the CPI mentioned earlier.[21] In Israel, after twenty years of full indexing of both social-security payments and treasury bonds the linkage was reduced to 70% in the late 1970s. During the twenty years of full indexation several significant devaluations occurred which have led to overestimations of the index to which the transfer payments and the bonds were linked.

What have governments done to rectify the situation? In effect they have chosen among four strategies to reduce the pressure on expenditures of full indexation of transfers:
— maintaining full indexing and reducing spending in other areas;
— raising taxes, borrowing more, or accommodating spending by increasing the money supply;
— revising the index;
— suspending full indexing.

Most governments have chosen to revise the index, but some have suspended or abandoned the practice through legislation.[22] For example, after the OPEC oil embargo in 1973 Denmark and the Netherlands suspended indexations; Norway decided to reduce the percentage of the CPI that was compensated for; Belgium and Israel revised the index used in COLA agreements and business contracts, as did Finland and France. In Finland and Israel, as well, employers and employees agreed to eliminate from the price index to which wages were linked the estimated effects of both changes in indirect taxes and significant changes in import prices, while Canada suspended indexing in 1975 as part of its anti-inflation program.[23] Some of these countries also eliminated fractions of the increases in the CPI from the index as a way of compensating for the effects of inaccuracies. Until 1972 Belgium eliminated 2% of the rise in the CPI from the COLAs. In

Israel, the automatic protections were more complex, but they, too, have always had "caps" and "triggers" in addition to eliminating the effects of changes in some prices from the index.[24]

It is not possible to prescribe which option governments should pursue, once full indexation to the CPI is practiced, for the important reason that the *change of the index* transforms indexation into an issue of wealth distribution and therefore of political choice. It becomes such an issue because the relative price changes that increased the index accompanied a decrease in the real value of the national income relative to the expected one. Full indexation thus automatically redistributed wealth from the nonindexed to those with full indexation. The choices governments face can be illustrated with the following example. If the Canadian nonretired public believes that the purchasing power of the retired population should be maintained at its historical level during periods of inflation and frustrated expectations, then they will agree to leave the indexation practice unaltered and either pay higher taxes or reallocate government expenditures in the retired people's favor. In effect, they agree to lower their own income expectations to protect the purchasing power of the elderly. In this case the government should have no difficulty in choosing either the first or the second strategy for continuing the practice of indexation.

But the same public may view the dilemma differently. The nonretired public may think that they did not make a commitment to protect the incomes of retired people at their historical level. Let us say this wealth redistribution occurred inadvertently because *no* attention was paid to the index to which social security was linked. The index overestimated the change in purchasing power and *increased* social-security benefits relative to wages. In this case the nonretired public will be against maintaining indexed benefits at their promised levels and against increasing taxes to pay for the rise in social-security benefits, and the government will encounter difficulties if it wants to maintain the practice of full indexation. But, due to the political process involved, the government may also find it difficult to revise indexation practices. This is unfortunate, since during the political debates much harm can be inflicted on the economy because of the rising deficits, increased borrowing, inflationary financing, and, in general, the increased uncertainty as to the government's future course of action.

Recently Stanley Fischer (1981) examined evidence from several countries on the effects of indexation on the inflationary impact of the oil-price shock of 1973. His data suggest that while wage indexation did not, in general, increase the inflationary impact of the oil shock, this impact was significantly stronger in some countries where governments had issued fully indexed treasury bonds. This finding is consistent with the analysis outlined above and suggests that the slow process of revising the practice of indexation has the harmful effect of increasing the rate of inflation. Fischer's results suggest that in particular, it seemed politically difficult to revise the index to which the assets of governments were linked retroactively, and governments tended to decide on an inflationary financing of their increased deficits. In Israel, in the late seventies, full indexation of social-security benefits and treasury bonds to the CPI has been revised, and now they are both only partially indexed. The revision should have been carried out earlier since for twenty years the effects of significant and frequent devaluations have not been eliminated from the index to which the benefits and the bonds were linked. The damage was probably done, and the more than 100% inflation of recent years can be partially attributed to the unexpected increases in deficits due, in part, to the inaccurate indexation practice and the decision to make up these deficits by inflationary financing.[25]

In other countries the policy has been, once it is realized that a distorted index has been chosen for indexation purposes, to revise the index rather than choose any of the alternative policies presented in the previous paragraphs. By revising the index the redistributive effects of indexation may be eliminated once and for all. In contrast, if transfers are temporarily diminished, taxes are raised, or the structure of government expenditure is altered, governments continue to run the risk of unanticipated deficits and the redistributional consequences outlined in the previous paragraphs. Of course, if the public decides to do so, it can maintain the purchasing power of some groups at their historical level. However, the decision to do so will then be explicitly stated by raising taxes and will not be due to full indexation to the CPI, a mechanical device that might have been adopted and used without much thought.[26]

What Are the Remedies?

Indexation of transfers has been undertaken in Canada to protect the purchasing power of low-income groups automatically and avoid the often lengthy legislative actions required in an inflationary environment to revise these payments to keep pace with inflation. The index chosen, however, has been the CPI. Little attention has been paid in Canada or elsewhere to the index to which payments were linked when the indexation practice was introduced. It appears that the redistributive consequences of such a choice were not anticipated. In most cases, however, the resulting arbitrary redistribution has turned the issue of indexation into a political one of how should the practice be revised.

Since no studies have been done on just how much Canada's CPI overstates the erosion of purchasing power by inflation, it is not possible to estimate the redistributive impact of both full indexation and use of the CPI. A rough estimate is possible, however, which suggests that the problem is severe. In 1976 Old Age Security, Guaranteed Income Supplements, and Spouses' Allowances cost $4 billion; Family Allowances cost $2 billion. Suppose that in a subsequent year 5 percentage points of the change in the

CPI was due to the measurement problems discussed above. That implies that if no compensation had been given for this 5 percentage-point increase in the CPI, $300 million could have been saved in the government's expenditures in that year alone, rather than being arbitrarily redistributed.

The unintended redistributive effects of full indexation can be prevented by one of the following methods:
— partial indexation—either by eliminating certain price changes from the CPI or by revising the index;
— indexation to changes in private-sector wages.
Let us discuss these two options.[27]

Partial Indexation. This option implicitly assumes that an indexation policy should take account of the sources of increases in the CPI. When the prices of some commodities rise, such as oil prices after the OPEC embargo in 1973, or food prices after a bad harvest, or import prices after an exchange-rate depreciation, not everyone can be protected without perpetuating inflation, widening government budget deficits, and creating redistributive consequences. These price increases are therefore candidates for removal from the CPI. The result of such a move to partial indexation means that everyone would then be protected, approximately, only against effects of changes in the money supply. As shown in tables 6.1 and 6.2, a close relationship exists between increasingly expansionary monetary policy and high and variable inflation. The point of protecting only against increases from this source is that such protection would have no redistributive effects. But this goal makes sense only if the monetary changes have indeed been accidental and did not come in order to redistribute wealth.

Another way to establish an index that would not have the redistributive effects of the CPI is to revise the index arbitrarily to allow for less than full protection, and index for only 70% or 80% of the change in the CPI.

The benefit of revising the index by eliminating the effects of some price increases from it is that the changes in the price index may then be more accurately measured. Its drawbacks are twofold: (1) it is difficult to reestimate consumption patterns frequently; and (2) the index can easily turn out to be subject to political manipulation. In this case people would cease to give much confidence to price indexes in general. This latter possibility is no mere speculation: such manipulation did occur in both Brazil and France, where arbitrary large weights have been given in the CPI to goods whose prices were controlled by the governments.

The benefits of having only a fixed 10% or 40% partial indexation are twofold: (1) it may save the time devoted to legislative actions for adjusting the payments and the tax brackets; and (2) it eliminates the lags for the beneficiaries during which their purchasing power would have diminished. The drawback of this method is that the percentage of partial indexation is rather arbitrary and may be frequently challenged.

Indexing to Changes in Private-Sector Wages. Indexing government payments to changes in wages in the private sector rather than to price indexes might alleviate some of the difficulties that have arisen. The benefits of this method come to the fore when the CPI rises but the productivity either stays constant or diminishes.

If transfer payments are indexed to this measure, their purchasing power will change in proportion to the purchasing power of wages. That is, if wages rise faster than the CPI, indicating rising productivity, benefits to recipients of transfers will increase by the same percentage as wages. And the contrary: if wages rise by less than expected, indicating diminishing productivity, the benefits of the recipients of transfers will rise by the same percentage as productivity, *whatever* happens to the CPI. Thus the distribution of incomes between recipients of transfers and wage earners will not be arbitrarily changed. In contrast, if government payments are indexed to the CPI, and the CPI rises when productivity and profits do not, the income distribution will be arbitrarily changed; the undesirable destabilizing effects of this change have been discussed.

The benefits of this method can be illustrated by a numerical example that has already been used. The unexpected 25% increase in energy prices has been shown to lead to a 3.75% rise in the CPI. Since neither productivity nor profits have changed (although profits in some sectors could have increased), assume that wages, on average, have not changed either. This implies that the purchasing power of wages as measured by the CPI has diminished by 3.75%. If transfer payments are linked to changes in wages, they will be left unchanged, too. In contrast, automatic indexation to the CPI would have increased these payments by 3.75%, leading to an unexpected increase in deficits.

Indexing tax brackets to this measure has similar benefits. If the CPI rises by 3.75%, but wages do not rise, the tax brackets will be left unchanged, and people will not automatically pay lower taxes. Thus revenues will not be unexpectedly diminished.

In conclusion, the suggestion that transfer payments and tax brackets be indexed to changes in wages in the private sector during inflationary periods does not imply that, if the public wants, government cannot change the tax brackets or increase transfer payments when the CPI rises but productivity and profits do not, thus redistributing wealth to some groups. This strategy of indexing only implies that these changes will not occur automatically in these circumstances.

3. Concluding Comments

The arguments and evidence presented in this chapter show how the traditional approaches to inflation can be combined with the one presented in the previous chapter to shed light on recent events—this may be its methodological value.

Its practical value is that it leads to a greater understanding of the problems that arise with indexation and inflation. The discussion on what index to choose, how to adjust it, how frequently to recalculate it, how to change monetary and fiscal policies if an inappropriate index has been chosen, and so forth, showed that indexation is no panacea. Since the evidence enables one to infer that indexation is more likely to emerge when there is greater monetary instability, policy makers should take into account these costs of practicing indexation together with the other costs of having an unstable monetary policy, when the decision is made to pursue such a policy.

The international comparisons in this study should help to illustrate two things: first, that there are lessons about the practice of indexation to be learned from the experiences and decisions made in other countries; second, that other governments, too, have adopted indexation under various political pressures without any consistent goal in mind. In some countries treasury bonds and social-security benefits have been fully indexed, but not taxes (Israel); in others, taxes and social-security benefits have been indexed but not treasury bonds (Canada); in still others, only social-security benefits, but neither taxes nor bonds (the United States). Moreover, in some countries while treasury bonds were indexed, the loans given by the government were not (Israel), which aggravated the problem of unexpected increases in deficits when inflation rates climbed. No rationale can be found for this variety of inconsistent practices, just as no serious discussion can be found on the negative effects of linking government transfers to the CPI, or on the decision to index them fully. Thus this international evidence shows not only how randomly we learn to adjust to new circumstances (evidence quite consistent with the views presented in chapter 1) but also that all the practical problems that appeared with indexation have been linked to the issue of the redistribution of wealth. This is to be expected, since it is this issue that is on people's minds (literally speaking).

Table 6.1 Average Rates of Change

	CPI	GNP Deflator	M1[a]
1949–52	4.6	5.5	—
1953–65	1.3	1.5	4.1
1966–71	3.6	4.6	7.2
1972–80	8.6	9.3	10.2

Source: Statistics Canada, CANSIM, Matrixes: 529, 531, 674, 7,000. Bank of Canada Data Series, no. B2013.

[a]Narrow definition of money: currency and demand deposits

Table 6.2 Average Rates of Change

Year	CPI	M1	Year	CPI	M1
1949	3.1	—	1966	3.7	7.0
1950	2.9	—	1967	3.6	9.7
1951	10.0	—	1968	4.0	4.5
1952	2.3	—	1969	4.5	7.0
1953	−1.0	—	1970	3.3	2.3
1954	0.6	0.2	1971	2.8	12.6
1955	0.1	9.1	1972	4.8	14.2
1956	1.4	2.0	1973	7.6	14.6
1957	3.1	−0.2	1974	10.8	9.4
1958	2.6	9.6	1975	10.8	13.6
1959	1.0	2.5	1976	7.5	8.3
1960	1.3	1.2	1977	8.0	8.3
1961	0.8	5.2	1978	8.9	10.0
1962	1.2	3.3	1979	9.1	7.1
1963	1.7	5.8	1980	10.1	6.4
1964	1.8	5.1			
1965	2.4	6.3			

Source: Statistics Canada, CANSIM, Matrixes: 529, 674, 7,000. Bank of Canada Data Series, no. B2013.

Table 6.3 Incidence of COLA Clauses within Collective Agreements and within the Total Canadian Private Sector

Year[a]	Collective agreements	Private sector
1951	21.4	—
1955	7.0	—
1962	2.6	—
1967	2.2	—
1969	4.5	9.2
1971	6.2	20.0
1975	33.9	53.9
1977	20.6	—

Sources: Jean-Michel Cousineau and Robert Lacroix (1981); David A. Wilton (1980), p. 5.

[a]The years were chosen in order to show the trends in the indexing practices. Regular 5-year intervals were either not available or could give a somewhat distorted picture of the trends.

Table 6.4   Contract Characteristics

			Average duration of contract		
Year	Percent of settlements with COLA	Percent of employees with COLA	with COLA	without COLA	Total
			(months)		
1970	3.0	5.2	34.1	25.7	26.2
1971	6.6	4.7	—	—	—
1972	14.5	35.9	43.9	23.7	31.0
1973	20.6	21.6	—	—	—
1974	36.9	39.6	20.4	21.2	20.9
1975	33.7	41.3	—	—	—
1976	26.3	38.3	34.8	15.7	23.0
1977	20.4	18.8	20.9	14.8	15.9

Source: Department of Finance, Canada, "Canada's Recent Inflation Experience" (1978), table 6. Original source: Collective Bargaining Division, Labour Canada.

[a]Data based on bargaining units of 600+, excluding construction.

Table 6.5   Realized Rates of Change in Wages

	All wages		Manufacturing sector only	
Year	Indexed contracts	Non-indexed	Indexed contracts	Non-indexed
1970	9.0	8.7	9.0	8.6
1971	10.8	8.2	10.2	8.7
1972	10.7	10.5	10.8	10.1
1973	12.3	12.5	11.9	13.4
1974	17.1	15.5	16.3	16.9
1975	16.2	18.5	16.3	17.9
1976	10.2	11.6	—	—
1977	9.1	8.6	—	—
Average	11.9	11.8	12.3	12.6

Source: Cousineau and Lacroix (1981), first two columns, p. 46; last two columns, p. 113.

Table 6.6   The Elasticity of Indexation Clauses in Quebec, Ontario, Canada, 1970–75, Public versus Private Sector

	Quebec		Ontario		Canada	
Year	public	private	public	private	public	private
1970	—	.36	—	.35	—	.31
1971	—	.30	—	.44	—	.39
1972	.64	.25	—	.32	—	.32
1973	.24	.28	.00	.48	.25	.38
1974	.00	.21	.23	.40	.25	.27
1975	—	.36	.06	.30	.03	.26

Source: Jean-Michel Cousineau and Robert Lacroix (1981), p. 112.

Appendix 6.1:
Indexation—Additional Viewpoints

In later years some papers on indexation have returned to the basic ideas outlined in the first section of this chapter.[28] On the other hand, monetarists and Keynesians have taken different approaches. This appendix discusses how indexation is more consistent with the monetarist approach than it is with the so-called Keynesian approach.

Advocating indexation in a monetarist framework is a consequence not only of its relation to the quantity theory but also of its view of government and its treatment of the labor market. The government imposes inflation as an unlegislated tax. Hence the purpose of indexing the bonds issued by the government is to force it to be "honest" in the sense that it would provide information as to the size of the government sector.[29] This argument appears, for example, in Friedman (1974b), and also in a later discussion on this subject from which the following quotations are taken:

> First I would like to see legislated indexing of governmental contracts in order to make the government honest. [Friedman 1974c, p. 2]

> Inflation is produced in Washington and nowhere else—it's produced by the government. [p. 11]

But Friedman discusses neither the practical problems linked to the inaccuracy of price indexes nor the fact that in some circumstances inflation may be viewed as a necessary tool to redistribute wealth. In fact these quotations view the government as an outsider to the economic system: "real" changes occur in the "economy," while monetary ones are caused by the capricious expenditures of governments.

Indexing in the labor market also seems consistent with the monetarist approach to this market.[30] In general models of wage indexation, the problem has been treated in the following way: when the wage contract is nominal, the change in wage rate is due to expectations as to the future price level; while when the contracts are indexed, such expectations are excluded and an automatic adjustment is paid. Thus disequilibrium in the labor market, where nominal contracts are assumed to exist, arises because of the difference between the expected rate of inflation (which is reflected in the contract) and the realized one. This view of the labor market originates with the monetarist view of the inflationary process,[31] and suggests that the main problem arises from inappropriate expectations as to the future inflation rate. Thus indexation is viewed as the solution to both problems: forcing

governments to be "honest" and bringing labor markets closer to equilibrium.

From here we can already perceive that there is much less cause for
introducing the system of indexation when taking a "Keynesian" approach.
One distinction of this approach is the attention paid to different price levels
rather than one price level. But there are greater differences between the
two approaches in their discussions of the causes behind inflation. In
Keynes's theory and in the Keynesian approach we may find political
necessities behind inflation. One such example is the already-mentioned
process described in *A Tract on Monetary Reform* (1923). That the process
described there is still consistent with Keynes after the *General Theory* is
evident from the description of the inflationary process in "How to Pay for
the War" (1940). In this essay Keynes advocates the introduction of compulsory savings, which, after the war, would be taxed through a "capital levy."
The rationale for this tax is found in the description of the inflationary
process of the *Tract*. Only by reading it is it possible to understand Keynes's
anxiety about what will happen after the war and the rationale for this tax. In
the *Tract* inflation is a consequence of the war, because there is a need for a
redistribution of income between the rentiers or bond-holding class and the
working class. In Keynes's opinion, there is an efficient alternative for
preventing inflation—"the capital levy." This tax is deemed by him to be
impossible politically (thus the words "efficient" and "just" do not have
much meaning if the tax cannot be imposed). Keynes (1923) explains:

> There is a respectable and influential body of opinion, which repudiat-
> ing with vehemence the adoption of either expedient, fulminates alike
> against devaluations and levies, on the grounds that they infringe the
> untouchable sacredness of contracts. . . . Yet such persons by over-
> looking one of the greatest of all social principles, namely the fun-
> damental distinction between the right of the individual to repudiate
> contract and the right of the State to control vested interests, are the
> worst enemies of what they seek to preserve. [p. 56]

The quotation emphasizes an additional difference between the approach
it describes and the monetarist approach. While in the above-described
process, "the people" force the government to use the tool of inflation, in
the monetarist approach the contrary seems to hold. Therefore, those who
view inflation as a political necessity cannot (in some circumstances) recommend indexation indiscriminately.

7   The Choice

"That's where the big mistake was: no movement should have been permitted, since we could exist only in immobility. The more immobile immobility is the longer and surer its duration."
A courtier in Haile Selassie's court.
Quoted in Ryszard Kapuscinski:
The Emperor: Downfall of an Autocrat

Explaining why wars occur—and coming up with some uncomfortable, although not totally new, answers—is a difficult challenge. How seriously can my explanations be taken? I believe that the explanation offered in this book—in contrast with the views of many other writers who have come to similar conclusions (though often without much precision)[1]—provides illuminating insights not only into the causes of war but into everyday aspects of human behavior.

In order to reemphasize the relative advantage of the approach presented in this book, this chapter will first show how it explains many of the unsolved problems that some historians and some political scientists have found in the works of their fellows. Then, the relationship between wars and population is discussed, with special reference to the wars of the twentieth century. The final pages of the book briefly summarize my views as to how the world would look if the danger of wars was significantly diminished.

1. Stability and Change

Now in all states there are three elements; one class is very rich, another very poor, and a third is a mean . . . it will clearly be best to possess the gifts of fortune in moderation; for in that condition of life men are most ready to listen to reason. But he who greatly excels in beauty, strength, birth or wealth, or on the other hand who is very poor, or very weak, or very much disgraced, finds it difficult to follow reason.

Aristotle, *Politics*

In a recent book on wars, Robert Gilpin (1981) raises this question: why is the problem of political change, which is crucial for understanding wars, a relatively neglected topic? He suggests these answers: that all theories develop from the static to the dynamic, and that rigorous theories of human behavior are still in their first stage and thus not applicable to wars; that in recent years there has been a decline in "grand theories" (à la Morgenthau, Kaplan, Haas) and a rise in marginal ones, which have diverted attention from fundamental problems; that theories of international relations lack a rigorous basis; that some political scientists have given up all hope of finding a theory, since historical events are so complex; and finally, that for social scientists the idea of radical changes that threaten accepted values and interests is not an appealing one.

If these are indeed the reasons for the neglect, the approach presented in this and my previous book suggests a solution and shows that most of Gilpin's reasons may no longer justify the neglect. The theory presented here is dynamic and not static. Also, while it refers to individual behavior, I have shown both here and in my first book how it can become a "grand" theory, in the sense that it sheds light on "grand" events like the "rise of the West," anti-Semitic outbursts, and wars and their possible consequences, in spite of the fact that it starts from a simple analysis of what motivates individuals. This is, in fact, its strength; it is an approach consistent with Harsanyi's (1969, p. 532) recommendation that the problem of social change must be explained in terms of personal incentives for some people to change their behavior. The approach may also eliminate the paradox Moore (1968) mentions, namely, that "as the rate of social change has accelerated in the real world of experience, the scientific disciplines dealing with man's actions and products have tended to emphasize orderly interdependence and static continuity" (p. 365)—all the evidence discussed here refers to *disordered* interdependence and change.

Gilpin's statement that some political scientists have given up all hope of finding a theory seems strange—what are these scientists doing, then? Gilpin suggests an answer when he mentions the last reason for the neglect of a dynamic theory: namely, that the prejudiced approach of some scientists toward the subject of political change orients their researches. The rationale for this prejudiced approach in terms of the model presented in chapter 1 is this: scientists brought up on particular views may be as afraid of radical changes as any one else, since these changes may either destroy their reputation and wealth (if they hang on to their views while the public bets on the new ones) or force them to abandon their customary views—and the latter is not an easy choice to make. It is naïve (and inconsistent with the evidence)[2] to assume that all scientists are motivated by a desire to look for facts, and are more flexible than other people. Most of them, like the Mandarins of earlier times, are, like everybody else, creatures of habits of thought.[3]

The relationships among wealth, changes in its distribution, and creativity have been noted by many social scientists, although vaguely. According to Gibbon, Montesquieu, and Polybius, the prosperity generated by a political conquest leads to the loss of moral virtues and to eventual decline. Schumpeter believed that the success of capitalism (another undefined word) destroys the risk-taking entrepreneurs who are responsible for economic progress; they just live on the wealth accumulated in the past. He applied his views to shifts in the position of families in the class structure and stated that their relative position undergoes change not in such a way that the "big" ones grow bigger and the small ones smaller, but typically the other way round, and summarized the evidence that emerged along the lines of the American saying: "Three generations from shirtsleeves to shirtsleeves."[4] Modelski's (1978) and Cipolla's (1970) views are similar: Modelski writes that there is a "Buddenbrooks syndrome . . . one generation builds, the second consolidates and the third loses control" (p. 232), while Cipolla concludes that "In every good family there is the generation that builds up a fortune, the generation that holds on to it, and the generation that dissipates it.[5] In this respect societies differ little from families" (p. 12). But these views are biased, because they derive their conclusions looking only at societies in turmoil, and neglect looking at those—which we call "primitive"—that succeeded in maintaining their stability. How "primitive" societies accomplished that, and why previous observations on cyclical human behavior refer only to societies where the major shocks that occurred were beyond human control are questions that were examined in my first book.

Gilpin (1981) criticized Modelski's and Cipolla's views on other grounds: he wrote that while the idea of "cycles of war and peace is intellectually attractive, the difficulties of long-wave theories in politics as in economics is that no mechanism is known to exist that can explain them. Thus . . . [this idea] is no more convincing than the more elaborate sixteenth-century formulation of the idea: 'I have always heard it said that peace brings riches; riches bring pride; pride brings anger; anger brings war; war brings poverty; poverty brings humanity; humanity brings peace; peace, as I have said, brings riches, and so the world's affairs go round'" (p. 232). However, as shown, the views presented here suggest an explanation for this process.

Finally, a few further views on creativity should be mentioned: according to Toynbee, the impulse to growth is always provided by creative minorities, and, at times, Toynbee seems to link the emergence of these minorities to changes in the distribution of wealth.[6] In Marxist analysis, everything that bourgeois society builds is built to be torn down—"all that is solid melts into air"—and it is the class structure that at first facilitates and later retards innovations. Similar views were shared by some middle-of-the-nineteenth-century Russian populists who raised the possibility that the cultural creativity of the West, far from being a condition waiting for all mankind, is a Western peculiarity that originated in its desperate moral anarchy. These views seem, at times, consistent with mine, although there are big differences: in contrast to Marxists, who look only at "masses" and "production processes" and have a deterministic view of history, I have looked at individuals and their ideas, and have shown how chance shapes history. In fact, Marx seemed to believe that history is governed by laws which cannot be altered by individuals believing in one or another idea (isn't that strange, considering the effect his ideas have had on other people's minds?). According to him the inner experience to which men appeal to justify their ends does not reveal either moral or religious truth, but just gives rise to illusions, individual and collective.

Where does such a belief put Marx himself? Doesn't *he* belong to this category, then? Interestingly enough, his correspondence reveals that he perceived himself as being *outside* of this general law: Edmund Wilson noted that Marx used a curious language in his correspondence, speaking about himself as being "confined" (*"eingeschlossen"*) among men, as if he were set above the world of miserable mortals[7]—quite an arrogant attitude (but see Isaiah Berlin's explanation of it).[8]

Yet there are similarities between my views and his. It is clear from the statements made above that Marx did not believe in the power of "rational" arguments or the continuous amelioration of the human condition. In fact it was his view that men can achieve their full potential through misery and alienation—what he perceived as the contradictory character of "progress."[9] He believed, however, that it is possible to *understand* the process. In this sense there is no disagreement between our viewpoints: one can understand the past from my model (without this fact implying an ability to predict the future with much precision). Indeed, my views also imply that things must get worse (for some people) before they can get better; otherwise there is no incentive to deviate from the beaten path. But then again our views do differ: while Marx believes that such a process leads to some ideal situation, my views suggest that it merely reflects a transition from one form of order (or status quo) to another.

Marx perceives correctly that within every order the higher classes in the social hierarchy have the incentive to insure themselves by inventing ideas which would conceal the arbitrary character of that particular order by proposing, as Berlin puts it "a dogmatic code, a set of unintelligible mysteries expressed in high-sounding phrases, with which to confuse the feeble intelligences . . . and keep them in a state of blind obedience" (p. 29). (How did Voltaire say it: "Our priests are not what simple folk suppose; / Their learning is but our credulity"? [*Oedipe* 4.1]) However, our views diverge again as to the implications of this observation: I suggest that this process is inevitable, every order being characterized by inventing ideas which, if believed, could maintain an existing social order, while Marx believes in the existence of an ideal order.

There are further similarities and differences: Marx perceives the history of societies as the history of inventions: they alter both the relationships within a society and people's habits. Within my model inventions may also have this effect. Yet there is a fundamental difference: inventions within my model are *not* gifts, but they stem from some people's suddenly disordered lives and the perception of inequality. Such disorder may occur for a variety of reasons: it may be due not only to previous inventions, but also to changes that are not under human control (like sudden, large fluctuations in population because of a lack of immunization to diseases, changes in climate etc.—accidental events that are anathema to dogmatic Marxists). The effect of all these changes is to alter people's positions in the distribution of wealth suddenly and thus induce innovation.[10]

Yet to present a fairer picture of Marx's views, one must briefly point out the two major schools of thought from which they emerged. During Marx's time one of the most influential philosophers was Hegel, who viewed history as the development of the "Absolute Spirit" and who believed that during any period an identical spirit reflected itself in the thoughts and acts of people. Ludwig Feuerbach, whose works strongly influenced Marx, has pointed out that such beliefs are an empty tautology: what is the spirit of an age if not the sum of the acts that characterize it? His suggestion was that material conditions determine people's thinking and that acting and material distress cause them. Berlin (1939) summarizes his views as follows: "to seek solace in an immaterial ideal world of their own, albeit unconscious, invention, when as a reward for the unhappiness of their lives on earth, they would enjoy eternal bliss hereafter . . . [and] all 'ideologies' whether religious or secular are often an attempt to provide ideal compensation for real miseries" (p. 58). There are similarities between these views and mine. The difference again is in the implications: while I perceive no alternative to this process, Feuerbach implied that knowledge of physical laws can control these tendencies and enable people to make conscious adaptations. It is against the background of these two contrasting viewpoints that we should view the ideas of Marx.

The proposition that he intended to refute was that ideas have the decisive effect on the course of history, and he advocated replacing it with the alternative paradigm that social struggles had the decisive effect (with inventions being perceived as manna from heaven). In the light of my model, both guesses (call them, if you wish, "ideologies" or "philosophies") provide only partial pictures. If the two are combined, as my views suggest that they *can be*, the consequences seem at first sight quite surprising, since the structures built on both of them seem to collapse. First, the story that emerges can be told in simple, ordinary language; one can get rid of the cumbersome vocabulary used by both Marxists and some philosophers. It also becomes evident that the gamble on new ideas is due to the perception of inequality (but that one's eyes are opened only if one's way of life is suddenly disturbed), and that chance plays a role. Thus, while the model combines the two extreme viewpoints, most of their implications are lost.

In particular, the model sets a limit to what we can "know" and discover, the limit being that we will not be able to discover the meaning of probability—as it appears in the definition of "thinking." If we could, we would "know" without error. But this sentence is a play on words, since the word "knowing" can no longer be defined. So let us abandon these philosophical issues (I tried to avoid them both in my previous book and here) and concentrate on practical, pragmatic matters.

2. On Wars and Population

Humans live best when each has his place to stand, when each knows where he belongs in the scheme of things and what he may achieve. Destroy the place and you destroy the person.

Frank Herbert, *Heretics of Dune*

The basic concern that motivates individuals, according to my views, is whether or not they are outdone by their fellows, either on the domestic or the international scene. Thus the solution for preventing restless creativity, gambles on strategies of war in particular, is to prevent this leapfrogging process. Can it be done?

The views presented here and in my first book suggest an answer: governments should try to introduce a system of incentives that would keep the population stable from now on: indirectly, this policy would eventually maintain people's position in the distribution of wealth, domestic and international. Let us first examine how I arrived at this conclusion, and then briefly summarize evidence that has not been discussed until now.

As pointed out by numerous writers, primitive societies are characterized by a wide variety of customs, which have, however, one goal: to keep their population stable. We also know that in this society knowledge is memorized and repeated and is not innovative, and that a wide variety of customs exist which maintain the stability of the distribution of wealth.[11] Based on a wide range of evidence (archaeological, literary, and statistical), many writers have attributed the invention of agriculture in various gathering and hunting societies, and innovations in agriculture in general, to sudden population growth.[12] Further, the features of the Middle Ages have been attributed by other writers to Europe's suddenly diminished population (taking into account the military technology that has already been achieved), while the end of this age and the "rise of the West" have been attributed by others to a suddenly rising population and the resulting breakdown of customs.[13]

There has been one missing link in all the aforementioned views, namely, how exactly are fluctuations in population related to innovations? The views presented in chapter 1 (and in more detail in my first book) provide one answer: they show that when population suddenly increases, the kind of existence people have been accustomed to can no longer be maintained, the distribution of wealth changes, and people start gambling on new ideas. Thus what we today call "progress," or "evolution," seems, in the light of these views, to be an illusion, since many new institutions just substitute for the trust and customs that shaped human behavior when populations were much smaller and were maintained at stable levels, while innovations and discoveries are the result of people's suffering.[14]

According to these views, leapfrogging may occur when a nation's population suddenly grows and its people gamble on new ideas. Threats to the stability of international relationships may thus be due to one of the following scenarios: some nations may gamble on customs that keep their population stable, with the result that the society eventually ceases to innovate, while others may bet on innovations that may support its rising population. The latter may outdo the others in terms of technology, military in particular, and the disorder may start. But even if the latter society as well eventually gambles on customs that keep its population stable, it may be too late for preventing wars, since this society has already outdone other nations that may then feel threatened. An alternative scenario that may lead to war can unfold when within the nation whose population is growing (and whose domestic distribution of wealth is therefore changing), there may be such internal turmoil that its constituents may gamble on a war, rather than on technological innovations. Another scenario may be mentioned: the innovations of one nation (literacy, radio, TV) may disturb the status quo in another and have destabilizing effects.[15] In conclusion, the aforementioned changes in the internal and external distributions of wealth, direct or indirect consequences of population growth, increase the probability of the outbreak of wars.

These conclusions on leapfrogging and on the relationship between wars and population growth have been recently emphasized by Gilpin (1981) and McNeill (1982), without, however, providing a rigorous basis for their views. On leapfrogging Gilpin (1981), for example, wrote:

> . . . Britain began to take the lead in the technologies of the first phase of the Industrial Revolution—textiles, iron and steam power. The growth of the British economy and the relative decline of French power caused increasing disequilibrium between France's dominant position and her capacity to maintain it. Eventually this struggle between a declining France and a rising Great Britain gave rise to the wars during the period of the French Revolution and Napoleonic era. [p. 134]

While McNeill (1982) concluded:

> Certainly the way in which German power was countered by the Allies of World Wars I and II conformed in all the essentials to two earlier passages of European history: the two bouts of war that constrained Hapsburg power, 1567–1609 and 1618–48; and the more widely separated struggles that checked French preponderance, 1689–1714 and 1793–1815. In each of these cases, as in the years 1914–18 and 1939–45, a coalition of States took the field against the ruler of the day who seemed on the verge of establishing European hegemony. [p. 308]

McNeill provides further insights and directly links wars to the altered distribution of wealth, to population growth, and to the fact that under this pressure, traditional ways of life cannot be maintained. He thus describes the human attitudes in Western and Eastern Europe before World War I:

> To have a foreigner to hate and fear relieved potential combatants from hating and fearing neighbors closer at hand. This was profoundly reassuring to socialists and proletarians as well as to the propertied class . . . The fact that war enthusiasm was less apparent in eastern Europe [is due to the fact that] urbanization had affected a smaller percentage of the population in that region, where the peasant majority still sought to follow a traditional pattern of life. [p. 302]

The similarity between this description and the predictions of the model in chapter 1 are obvious. It should be also noted that Marc Ferro in *La Grande Guerre* and Emmanuel Todd in *Le fou et le prolétaire* suggest that because the artisan and shopkeeper classes were especially under pressure before 1914, they expressed their frustration by transferring hostility to a foreign enemy. Similar points are made by Wright (1964) in his massive study of wars:

> Rapidly changing technological and economic conditions which seriously alter the position of economic classes cause exceptional unrest. The development of commerce and industry deteriorated the relative position of landowners and peasants and caused much unrest in Renaissance Europe. World War I, inflation, and depression deteriorated the relative position of the middle class in much of Europe and caused serious unrest in the 1930's. The industrial revolution . . . produced extensive technological unemployment and violence in Chartist England. Similar conditions led to violence in China in the 1920's. [pp. 273–74]

Wright also notes the relationship between wars and population:

> Whereas . . . a country increasing in population less rapidly than its neighbor may view with increasing alarm the shift of the balance of power against it. These conditions, which were obvious in the relations of France and Germany from 1870 to 1890, may, of course, be altered by the establishment of alliances, as when France, with a stationary population, allied itself in 1891 with Russia whose population was growing more rapidly than that of Germany. Germany, which previously had viewed its relations with France with comparative equanimity, then became alarmed. [pp. 285–86]

On the relationship between wars and population, McNeill (1982) concludes:

> . . . if the democratic and industrial revolutions were, among other things, responses to population squeeze . . . towards the end of the eighteenth century, the military convulsions of the twentieth century

can be interpreted in the same way, as responses to collisions between population growth and limits set by traditional modes of rural life in Central and Eastern Europe in particular . . . a basic and fundamental disturbance to all existing social relationships set in whenever and wherever broods of peasant children grew to adulthood in villages where, when it came time for them to marry and assume adult roles, they could not get hold of enough land to live as their forefathers had done from time immemorial. In such circumstances, traditional ways of rural life came under unbearable strain. Family duties and moral imperatives of village custom could not be fulfilled. The only question was what form of revolutionary ideal would attract the frustrated young people. [p. 310]

while North (1981) points out that population seems to be a two-edged sword; at times a factor in domestic and international conflicts, and at other times inducing innovations.

If my views and those of these writers are accurate, wars can be prevented if populations are kept stable; what can governments do today in order to achieve this goal? First they must agree on a common objective—keep the population stable. If governments cannot agree, the domestic and international turmoil that has characterized the last centuries will continue. But if governments agree on this goal, each must try in its own way, taking into account the customs of its people, to formulate incentives to decrease the marriage rate, decrease fertility, regulate the age of marriage, encourage, or at least be tolerant toward, abortion and homosexual practices, allocate less to research which may increase life expectancy.[16] These adjustments are painful and may be difficult to enforce in democratic societies, in particular if large segments of these societies are still trapped by customs on which people gambled when increases in population were perceived as beneficial. (This may have been the case at times when, due to reasons beyond human control, the population suddenly, significantly decreased [see Brenner 1983b, chap. 3] and some influential groups believed in the superiority of one type of custom and one type of government. The mercantilists, for example, believed that what made a nation wealthy and powerful was a large number of poor who could be kept fully employed. Thus they strongly recommended policies to increase the population [from marriage to various tax incentives, including family allowances], hoping that these methods would provide them with the army of poor they perceived as beneficial. Some of these policies still survive in today's tax-exemption programs. If my views are correct they should be eliminated [gradually].)

But, as emphasized in chapter 1, even if our goal is "happiness" (that is, trying to provide a range of insurance that would minimize the effects of sudden fluctuations in people's lives), we cannot reach the conclusion that the solution for each form of government, for each society, should be the same. Yes, the ultimate, fundamental mechanism that motivates people everywhere and at all times does seem to be the same. But one must be aware of the variety of ideas that people have gambled on in the past that have shaped our lives today; what is good for one society with one set of customs is not necessarily good for societies that, accidentally, have gambled on others. Those who believe that customs and traditions should be and can be altered in a predictable fashion and the quest for a unique "ideal" solution pursued should read Europe's bloody history (on which we happen to have the most information) more closely and also look at recent histories, like that of the small tribe, the Iks.[17] When they were deprived of the mountains valleys of Northern Uganda where they had lived their quiet lives as nomadic hunters for who knows how many centuries, the consequences were disastrous: their customs broken, their conversation today consists only of ill-tempered demands. They share nothing. They do not sing. Children are turned out to forage as soon as they can walk, and the elderly are left to starve, the children snatching their food. Shouldn't one learn something about the importance of customs when looking at this unfortunate experience and be afraid of the consequences? Who knows how these, by our definition, "crazy" and desperate people will eventually behave?

But if leaders of governments would find the courage to pursue the goal of keeping their populations stable from now on, people would gamble on a wide range of innovations until customs defined a new level of wealth and the stability of the internal and external distribution of wealth was eventually maintained. During this process, insurance against abuses of power, which may significantly alter the distribution of wealth and forestall the plan's success, must be provided.

The final distribution of wealth that societies reach may not be egalitarian (it was not egalitarian in primitive societies either). But perhaps there is a good reason for that: were unfortunate shocks, beyond human control, to occur once the world settled at a stable state, the inequalities in the distribution of wealth would guarantee the survival of the human species. If the distribution was egalitarian, and an unfortunate shock occurred, people would be unable to break habits of thought, and would thus be unable to survive. While such a notion is controversial, it seems to follow from the views on human creativity presented in chapter 1 (the material presented in this and in my previous book provides, of course, only indirect evidence which does not contradict them).

How would the world look if nonegalitarian stability were achieved and no significant shocks beyond human control occurred? Habits would guide human lives, and children would be encouraged to memorize and to imitate, but not to innovate. If, by chance, some people had fluctuating lives, and the entrepreneurial trait surfaced, either social customs or laws would turn out to be sufficiently discouraging to induce them to have second and third thoughts about carrying out entrepreneurial acts.[18] Arts would become repetitive (like the famous Chinese paintings were for hundreds of years), and symmetry in paintings, music, and literature would probably be admired.[19] At the same time, the painful creativity symbolized by the legend of Prometheus and by Beethoven's music would probably be looked upon with some sadness mixed with irony, as an expression of the curious excursions of the human mind under stress. To *us* such a world might seem boring: gone would be the ideal that music should arouse passions, and the revolutionary ardor of Druids versus Romans (in Bellini's *Norma*), of persecuted Protestants against royalists (in Bellini's *Puritani* and Meyerbeer's *Huguenots*), of Jews against Egyptians and fate (in Rossini's *Moses*) and against Babylonians (in Verdi's *Nabucco*), of Swiss against Austrians (in Rossini's *William Tell*), of outsiders against established order and fate (in Verdi's *Ernani* and *Il Trovatore*)—all written during the nineteenth century when nationalistic fervor swept Europe—would probably seem artificial, dissonant, and noisy.[20] Instead, the disciplined style of submitting melody to a repetitive text will probably become the "new" ideal, and the laments of people who accept their fates, such as those found in the repetitive music of Purcell, Monteverdi, and Handel, may again serve to express human sentiments.

The new trends will probably be represented by musicians like Glenn Gould. In a profile of Leopold Stokowski he wrote that the nineteenth-century piano concertos (and all the concertos of the romantic era) with their dramatic confrontation between the orchestral collective and the pianistic individual were "immoral." So he concentrated on Bach. Recently Peter Shaffer (1984), commenting on Mozart's music, wrote: "Hearing the [Concerto in A Major] now, I still become amazed all over again by its *certainty*. The best of Mozart's works . . . demonstrate the thrilling paradox at the heart of created things: They justify obedience to form . . . They celebrate the idea of the correct, and prove beyond dispute the necessity of artifice . . . There is, in fact, something almost terrifying about this restraint. Over and over as one listens, a joyful shadow, a shadowed joy, seems to pass swiftly over the music . . . It makes the rigorous turbulence of Beethoven seem overinsistent, and the lachrymose iterance of Mahler largely hysterical" (pp. 23, 27).

There is nothing terribly novel about these observations: they have been made from the time of Aristotle, through that of Adam Smith, to that of Anatole France. Aristotle objected to Hippodamus' proposal for awards for new ideas: in a *settled* regime he considered them suspect. For "a citizen will receive less benefit from a change in the law than damage from becoming accustomed to disobey authority . . . the law itself has no power to secure obedience save the power of custom, and that takes a long time to become effective" (*Politics*, pp. 82–83). And as Smith (1795) puts it in one of his lesser-known essays (*The History of Astronomy*) men "have seldom had the curiosity to inquire by what process of intermediate events" a change is brought about, where "the passage of the thought from . . . one object to the other is by custom become quite smooth and easy" (pp. 44–45), and "it is well known that custom deadens the vivacity of both pain and pleasure, abates the grief we should feel for the one, and weakens the joy we should derive from the other. The pain is supported without agony, and the pleasure enjoyed without rapture: because custom and the frequent repetition of any object comes at last to form and build the mind or organ to that habitual mood and disposition which fits them to receive its impression, without undergoing any violent change" (p. 37). In such circumstances the number of ambitious, strong leaders who will emerge from within the military in particular is relatively small. In *The Revolt of the Angels*, Anatole France (1953) probably provides an accurate description of why this might be the case: "In Ialdabaoth's army, happily for us, the officers obtain their post by seniority. This being the case, there is little likelihood of the command falling into the hands of a military genius, for men are not made leaders by prolonged habits of obedience, and close attention to minutiae is not a good apprenticeship for the evolution of vast plans of campaign" (p. 206).

Are these features disappointing? Not necessarily, since they describe our choice:[21] we can decide either to look at the virtues of fluctuating populations and wealth distributions, and live with their consequence in spite of their dark sides, or to choose stability and both remember and teach our children (in Barton-type sermons, as in Bellamy's (1887) *Looking Backward 2000–1887*) the old curse: "You should live during interesting times!"

This choice is ours.

Notes

Chapter 1

1. The summary presented below is based mainly on Organski and Kugler (1980), Keynes (1919), and Waltz (1954). See also Morgenthau (1960) and Choucri and North (1974).

2. Notice that in the first sentence of this quotation Organski refers to a pyramidal power distribution, a distribution that plays a decisive role in my formal model.

3. For the definition of "growth" in my model see Brenner (1983b), chapters 2 and 7.

4. See appendix 1.2, especially Schumpeter's opinion on this point.

5. The word "suddenly" implicitly implies that an event happened which was not taken into account by the individual when he was forming his expectations.

6. See chapters 1 and 2 in Brenner (1983b).

7. But stability may be maintained if an idea (invented elsewhere) is adopted and benefits from it are redistributed without disturbing the already-achieved order.

8. An act (killing another person, for example) may be viewed as criminal when committed within a nation, but may be called "terrorist" if committed by one or more individuals from another nation.

9. Or, as Schumpeter (1919) wrote, nationalism "satisfied the need for surrender to a concrete and familiar superpersonal cause, the need for self-glorification and violent self-assertion" (p. 14).

10. Schumpeter (1919), for example, notices that the Roman Empire's imperialism was different from the one he defines since "the policy of the Empire was directed only toward its *preservation*" (p. 65. italics added).

11. For more on what "wealth" means, and why the term must include insurance based on customs see Brenner (1983b), chapters 2 and 3, where the relationship between individual and aggregate perceptions of "wealth" is also pointed out. See also chapter 7 in that book on this issue.

12. Clausewitz (1832), for example, noted that "the necessity of fighting very soon led men to special inventions to turn the advantage in it in their own favor" (p. 171).

13. Waltz (1954), for example, wrote: "War most often promotes the internal unity of each state involved. The state plagued by internal strife may then, instead of waiting for the accidental attack, seek the war that will bring internal peace. Bodin saw this clearly, for he concludes that 'the best way of preserving a state, and guaranteeing it against sedition, rebellion, and civil war is to keep the subjects in amity one with another, and to this end, to find an enemy against whom they can make common cause.'" (p. 81). Similar points were made by Waelder (1967), who quotes Halevy (1930) who, examining World War I, "suggested that the German leaders were motivated consciously or unconsciously by fear of a revolution and by the idea of uniting the nation through a national war. They hoped to strengthen their

own prestige and to secure their power by what was confidently expected to be a short and victorious war" (p. 275). Waelder also notes that Halévy's interpretation seems the same as that given for Bismarck's policies in 1860: then, too, revolution was feared by many (see chap. 2 on one of Bismarck's policies). Also, as Hitler's propaganda put it: "The proportions between space and people have now been reversed. The problem of how to feed a great people in a narrow space [i.e., the Germans] has changed into that of the best way of exploiting the conquered spaces with the limited number of people available" (*Die Wirtschaftskurve* no. 4 [November 1941], p. 272)—this problem appeared following the gamble of some Germans on Hitler's idea of finding new spaces for them, a gamble taken after a relatively long period of internal turmoil—more about the Germans' gamble later in the chapter.

14. Other social scientists have expressed views on the points discussed in this section. Gilpin (1981) wrote: "As the flow of tribute or economic growth slows, however, the conflict over relative shares of the economic surplus intensifies, despite its detrimental consequences for the overall welfare of society. As a result, periods of decline tend to be characterized by exacerbation of internal social and political conflicts, which in turn further weaken the society" (p. 137). And McNeill (1982) wrote: "To have a foreigner to hate and fear relieved potential combatants from hating and fearing neighbours closer at hand. This was profoundly reassuring to socialists and proletarians as well as to the propertied class" (p. 307). As to the uncertainty concerning the ideas that people may gamble on, it is useful to point out that McNeill makes a comparison between the universalist ideology the French gambled on in the nineteenth century and the principles that the Germans have gambled on during the twentieth century ("the Germans' bid for European hegemony was fought in the name of narrowly exclusive, nationalist, and racist principles," p. 313). One can point out, however, that the Versailles treaty may have contributed to the different attitude of the Germans—see the discussion in the text on the implications of the treaty. As to the relationship between wealth and population (expressed in the quotation in the text) see Brenner (1983b), chapter 2. The fact that a *deteriorating* situation leads people to rebel or fight wars has been noted, in fact, since ancient times. Aristotle in the *Politics* wrote that "in striving to avoid loss of money and loss of status whether for their friends' sake or for their own, men often bring about revolutions in their states" (p. 193), while rabbis discussing the Bar Kochba revolt suggested that "the revolt was not planned; rather it was the result of a deteriorating situation. The populace was not united. It was split by bloody internal controversies—a situation summarily described by the term 'causeless hatred'; it was this, the rabbis said, that brought on the destruction" (Harkabi, 1983, p. 6). On Plato's views and on more of those of Aristotle and some rabbis see appendix 1.3. Very recent observations are similar. Stephen Kinzer (1983), in a *New York Times* article entitled "Brasil Fearful of Explosion of Discontent," quotes Antonio Rangel, a Brazilian political scientist, on this point: "Poverty alone does not explain the explosive nature of the crisis, because poverty has always been serious here. The problem is 'pauperization,' the phenomenon of people slipping down the economic ladder. They are getting progressively poorer and that creates confusion and resentment" (p. 16).

15. This summary is based on Organski and Kugler (1980), pp. 106–7.

16. Without presenting a formal model, Olson (1982) makes a similar point.

17. It should be noted that one cannot say whether or not the views on which the populace gambles are "good" or "bad." But Harkabi's (1983) observation on this point should be noted: "This distinction between leaders who may mislead and the people who are drawn in by them . . . may have the practical implication that once it becomes evident that the leader's ideas were incorrect, leading to disaster, they may expiate their guilt. This mechanism may be useful in maintaining a group's *mental health*" (p. 19; italics added). See also Jung (1964) for an examination of the German character and his interpretation of its relation to feelings of inferiority. See appendix 1.1 on "mental health."

18. For more about Thucydides' work see Gilpin (1981).

19. There is an enormous literature on Machiavelli that I do not intend to survey. Some essays agree, while others disagree, with Machiavelli's views, yet none of them has the benefit of a precise Machiavellian model to nail down its arguments. A beautifully and very clearly written summary of the views of many writers, however, can be found in Berlin (1981). For a model explaining conservatism, but not departures from it, see Kuran (1984), who links it with some of Machiavelli's views.

20. While Berlin mentions many other reasons for the attacks on Machiavelli and the attention his works drew (one being his clear language and his avoidance of categorizing his views in the technical language of one of the existing disciplines) he, too, reaches the conclusion that the fact that there is no "ideal" society may have been the idea that most disturbed both his contemporaries and later generations. Among them there were many who believed in a unique solution that could benefit all mankind. If my views and those of Machiavelli are correct no such solution exists.

21. In particular, once customs are maintained and a regime established, inequality becomes legitimized by traditions (based on religious or biological or other ideas) and is therefore not viewed as a burden.

22. Marc Bloch (1953) noted that "The first tool needed by any analysis is an appropriate language; a language capable of describing the precise outlines of the facts, while preserving the necessary flexibility to adapt itself to further discoveries and, above all, a language which is neither vacillating nor ambiguous" (p. 157). Such a language must be morally neutral—the language of my model seems to be.

23. See Holdsworth (1903), p. 406.

24. Clausewitz wrote:

"War is a game . . . therefore, from the commencement, the absolute, the mathematical as it is called, nowhere finds any sure basis in the calculations in the Art of War; and [that] from the outset there is a play of possibilities, probabilities, good and bad luck, which spreads about with all the coarse and fine threads of its web, and makes War of all branches of human activity the most like a gambling game . . . Theory must also take into account the human element; it must accord a place to courage, to boldness, even to rashness. The Art of War has to deal with living and with moral forces, the consequence of which is that it can never attain the absolute and positive. There is therefore everywhere a margin for the accidental . . . [pp. 116–77]

As Waelder (1967) notes, the theory that viewed wars as a by-product of the monarchial system (and which was based on Clausewitz's powerful views) appeared so convincing that as late as 1918 President Wilson demanded that the "king of Prussia" be deprived of the power to declare war. But when it turned out that wars

202   Notes to Pages 19–27

were fought even after the disappearance or loss of power of monarchs, the theory was abandoned.

25. Schumpeter's views bear a strong similarity to those of Saint-Simon. The latter saw the historical process as a continuous conflict between economic classes. According to him the ruling classes are seldom flexible enough to adapt themselves to new circumstances. Eventually, the lower classes will try to overthrow them, although once they succeed and a new order is established the same process will repeat itself. Thus, according to Saint-Simon, human history is the history of such conflicts.

26. The mechanism suggested by the model presented in this chapter enables us to make this link.

27. Elaborate discussion of the relationship between "growth" and population appears in Brenner (1983b).

28. See McNeill (1982) and Brenner (1983b).

29. For more evidence on primitive societies, see Brenner (1983b), chapter 2. On the relationship among gambling, war, and religion in such societies see Huizinga (1944), whose work is quoted in appendix 1.1.

30. After this book was finished, *Time* (25 October 1982, pp. 65–66) reviewed a book written by a British diplomat, Robin Renwick, entitled *Economic Sanctions*. Renwick's study shows that while sanctions can seriously damage the economy of target nations, these nations have displayed impressive adaptability (and not only because of the difficulty of imposing sanctions effectively). Renwick argues that sanctions have failed to achieve their political goals and in fact have "stiffened" the behavior of the sanctioned nations. In spite of the lack of success of such policies, Renwick argues, they serve a useful purpose, since they are often the only way, short of war, for some nations to express their outrage, and they may deter other nations from violating the international code of behavior. In the context of this chapter, the main point to note is the sanctioned nations' adaptation to the worsened conditions (when all the population, or a large fraction of it, stood behind its sanctioned leadership).

31. There have been other approaches to examining wars, but they are either dogmatic or sterile, leaving out all discussion of human behavior. For example, in contrast to Clausewitz's views in which "the State" has interests, in the Leninist image the classes have interests and the ruling class uses the State to promote its own. According to Lenin, whose philosophy is rooted in Marx's writings, the dynamics of a capitalist economy with its permanently increasing demand for new labor and new materials is the root of war since it leads to competition for them among the capitalist states. Game theory, which has been frequently used in academic discussions on wars, seems totally irrelevant. For several reasons: not only does it rely on an implausible definition of "rationality" (see Rapoport's 1982 criticism), and not only must the probabilities for all outcomes be listed (!), but attitudes toward risk-taking are absent! Indeed, Rapoport notices that when risk-taking is important (when is it not?), strategic competence is of no help (p. 423). Schelling (1958), too, has emphasized the irrelevance of zero-sum games as paradigms of conflict. Another criticism of models based on game theory is that they assume that players have a clear idea of *all* contingencies. But since during wars there is nothing so rare as a plan (as Napoleon put it), applying such models in order to say something on wars seems quite questionable.

32. Although, needless to say, many writers have emphasized the role personalities play in the process leading up to wars and during wars—from Aristotle, who wrote that "For the efficiency of an army consists partly in the order and partly in the general; but chiefly in the latter, because he does not depend upon the order, but the order depends upon him" (*Metaphysics*), to Stoessinger (1978), and all those who wrote biographies of Napoleon, Hitler, Stalin, etc.

33. A similar point has recently been made by Hirshleifer (1977). Although few economists have admitted bluntly that what has been called "the economic motive" may be unimportant, Schumpeter (1919) criticized explanations of wars in terms of either concrete advantages or personal drives for power. He suggested that one must seek an explanation by realizing that it is not what economists call "rational ends" that motivate human behavior. Rather, the fundamental motive arises from impulses that, from an economic viewpoint, are nonrational. Knight's (1947) views were similar. In *Freedom and Reform* he says that ultimately the economic factor is relatively superficial and unimportant, and that conflicting economic interests are relatively unimportant as a cause of war.

34. The theory of risk presented here, based on Brenner (1983b), is different from the one used in economics (based on Friedman and Savage 1948 and Arrow 1970), in which attitudes toward risks depend on people's "taste," and thus no predictions can be derived as to who is more or less likely to take risks. For a criticism of that model see Brenner (1983b). For other criticism see Alchian (1953) and Hirshleifer (1977).

35. Notice that the mathematical presentation assumes in all cases examined the existence of either customs or laws that lead one to expect the protection of property (as defined by those customs or laws). Also notice that there is a chicken-and-egg problem within this model, a problem I do not intend to deal with. The perception of risk in this model appears because an individual is already assumed to belong to a group (the model works even if that group is defined as just the family). But how and why the *first* group of people was formed, and how they arrived at their customs this model cannot say.

36. The justification for this approximation is that the redefinition of the utility function (which now implies gambles on new ideas and, as a consequence discontinuity) requires a redefinition of "wealth" which implies that only broad categories of wealth can be identified. (Notice two things, however: First, that later, when criminal and creative acts are examined, *no* approximations are made. Second, that the approximation may not be greater in magnitude than the one made in the neoclassical analysis in which the effects of individuals and of firms are neglected in the analysis of "perfectly competitive makets.") One practical implication of the redefinition is that categories of wealth can be identified by symbols of prosperity: which neighborhood one lives in, how many maids the family has, and not whether or not the family dinner consists of going two or three times per week to MacDonald's. (See also Schumpeter 1927 on this point: he argues that intermarriage among members of the same class is such a symbol.) This redefinition of the utility function (and of wealth) implies a more tolerant behavior than the neoclassical approach allows. An example: while the latter cannot make sense of the stockbrokers' recommendation to "invest as much money as you can afford to lose," this model shows that this statement makes sense since the words "afford to lose" refer to one's standard of living in a certain wealth category. It implies that one should invest as

much as one can afford to lose without falling to the category of a lower class, which would induce one to get involved in that most difficult of acts: thinking. The model defines this word simply: as gambling on novel ideas.

37. The evidence on gambling appears in Brenner (1983b), Brenner and Brenner (1981, 1983), Gabrielle Brenner (1983), and chapters 2 and 3 in this book.

38. It is useful to note that in my methodology I have followed Coase's (1937) suggestion that economists explain the *emergence* of some markets rather than deal with existing structures and "explain" them by characteristics of the utility function. For the latter is, in fact, no explanation at all. Coase (1974) also suggested that one should examine the "market for ideas." Yet he does not mention that examining this issue means discussing how people *think*, and that if one had such a model, all aspects of human behavior should be examined in its light. For isn't "thinking" what human behavior is all about?

39. For rigorous examination of this prediction see the sources mentioned in note 37. But as anecdotal evidence consider the case of Fred Smith (the entrepreneur who created Federal Express). As he disclosed in a recent CBS "60 Minutes" program, he found himself a few years ago on the verge of bankruptcy. Though he was not a gambler (and did not gamble after that event), he then went to Las Vegas, where he won $27,000 which saved his company.

40. See Brenner (1983b) and Brenner and Brenner (1981). The idea that behavior is relative appears implicitly in the works of Duesenberry (1949) and Easterlin (1974). But neither of them related his arguments to attitudes toward risks or perceived that when one changes the model one *must* redefine the vocabulary.

41. It may be noted that the proofs are symmetrical for sudden increases in wealth: then one has an incentive to increase one's insurance.

42. Notice that in the model the entrepreneurial gambles start when an already achieved order is disturbed. When no such disturbances occur, many customs in the society can be viewed as ideas that people have gambled on in order to increase stability. Chapters 1 and 2 in Brenner (1983b) examine such customs in primitive societies. It may also be useful to point out Bateson's views on this issue: "In conventional economic theory it is assumed that the individuals will maximize value, while in schismogenic theory it was tacitly assumed that the individuals would maximize intangible, but still simple variables such as prestige, self-esteem, or even submissiveness. The Balinese, however, do not maximize such simple variables . . . In Balinese society—we find an entirely different state of affairs. Neither the individual nor the village is concerned to maximize any simple variable. Rather they would seem to be concerned to maximize something which we may call stability" (pp. 121–24). The model presented here shows that the goals of "maximization" and of "stability" can be compatible.

Chapter 2

1. See Ashton (1889) and Tec (1964).

2. For a history of state lotteries in the United States, see Weinstein and Deitch (1974), chapter 2.

3. See Johnson (1976) for a history of Canadian lotteries.

4. Staff document. Loto Québec.

5. Although one can "entertain" the thought of winning the big prize, the use of the word "entertain" does not imply that the act is just "entertainment" in the sense of merely bringing some pleasures of the moment.

6. More precisely: people who find themselves unexpectedly with a greater number of surviving children than they expected will tend to gamble more.

7. See Lazear and Michael (1980), for instance.

8. See Brenner and Brenner (1981), Kaplan (1978).

9. See Weinstein and Deitch (1974).

10. Further evidence comes from a survey done by the Massachusetts State Lottery Commission which found that: "the public prefers a game offering a relatively small number of large prizes to one offering a single very large prize and a great many very small ones. It overwhelmingly favors a single top prize of $100,000 over a large number of top prizes of $1,000 each."

"By 15 to 1, the public would rather have one chance in 1,000 of winning $300 than one chance in 2 of winning 50 cents" . . .

"The Massachusetts Lottery has no competition as the public's favorite form of betting. Betting at race tracks is a distant second. The 44% of bettors who prefer the lottery choose it because of its good chance of winning big prizes at low cost to play." Furthermore, 88% of the lottery-ticket buyers in Quebec polled in 1976 answered that their main reason for buying a lottery ticket was the chance of winning a big prize, and 80% answered that their main reason for buying a lottery ticket was to get rich. (These data appear in Robert Sylvestre's Marketing Study 1977).

11. These data are taken from Brenner and Brenner (1983).

12. The maximum Inter-Loto prize was then $250,000; of Super-Loto, $1,000,000; and of Loto-Perfecta, $100,000.

13. Thus, buying a lottery ticket represents a strategy for a person with less income to become richer and not, as usually asserted, an overestimation of the chance of winning.

14. While the same pattern exists for all three lotteries, Inter-Loto is obviously the most popular. This may be due to the lower cost of an Inter-Loto ticket ($2 a ticket versus $5 for Super-Loto and from $1 to $24 for Loto-Perfecta). Moreover, Inter-Loto and Super-Loto tickets are available at more outlets (16,000) than Loto-Perfecta tickets (only 2,000). All this may explain why Loto-Perfecta is less popular. It is also true that Loto-Perfecta is a "number" game, the other two being "pure" lotteries. This may influence the relative appeal of the three options.

15. It may be also noted that the possibility of winning big prizes by betting on soccer games depends on differences of opinion. If all gamblers shared the same views, and the outcomes turned out to be those expected, the gamble would cease to be interesting (and the soccer games, too).

16. The information on the Swedish gamblers is drawn from the tables presented in Tec (1964).

17. There is not enough information to enable us to infer whether this increased percentage is due to age or to higher income.

18. See Freud (1947).

19. See Freud (1942).

20. The writers seemed to neglect the fact that Freud admitted that his was a "trivial essay," "written reluctantly"—to quote Freud himself.

21. However, it should be noted that Ashton has two insights that are consistent with my model: he writes that "my objective is to draw attention to the fact, that the money motive increases, as chance predominates over skill" (p. 2;) and that "there is a class of gambling which is not considered harmless, but beneficial and even necessary. I mean insurance. Theoretically it is gambling proper" (p. 275). However,

Ashton does not perceive that the difference can be explained in a model in which people try to preserve, or enhance their position in the distribution of wealth. See the model in chapter 1 and in Brenner (1983b), chapter 1.

22. See chapter 5 in Tec (1964).

23. See Tec (1964), pp. 72–73.

24. See Tec (1964).

25. See Allen (1952), Tec (1964), and Brenner (1983b).

26. See Robert Sylvestre (1977).

27. For some similar arguments see Merton (1957).

28. It is clear that this is a prejudice since the facts contradict their opinions. People should thus be careful when listening to social scientists who provide no data to support their views. Their views might be biased.

29. See Tec (1964), p. 32.

30. Welfare payments, however generous, will never be able to accomplish that.

31. See the summary of arguments and of evidence in Brenner (1983b) in which the Canadian data are discussed. For data on the U.K. one may consult Carr-Hill and Stern (1979); on the U.S., Ehrlich (1974) (who always introduces a measure of inequality in his empirical tests), and Danziger and Wheeler (1975).

32. The reason for the *increased* crime rates is that we can say only that people whose position is worsened become more likely to commit crimes, but we do not know whether or not somebody who *became* richer did or did not commit crimes previously. The model makes no predictions on static circumstances. As to revolutions, see the sources quoted in Waelder (1967) and in Brenner (1983b), chapter 1.

33. How this process was carried out during the eighteenth century is discussed in the next chapter; for other centuries and for primitive societies, see the description of the process in Brenner (1983b). As to the observation that those who suddenly become poorer will tend to commit crimes, both the evidence mentioned in note 31 above and the almost daily observations made in magazines seem consistent with it. For example, in the *Economist*, 16 October 1982, the writer on "South Florida" attributes the more violent behavior of blacks to the fact that "unlike blacks in almost every other American city, they cannot even count on a monopoly of the low paid, low-grade jobs: Miami's blacks always live in danger of losing their jobs to the latest wave of immigrants off the islands prepared to work for peanuts" (pp. 21–22).

34. Current knowledge does not enable us to carry out any precise tests on this issue. Thus an allocation of one-third of the total resources devoted to crime prevention to each of the three methods is as good a starting guess as any other.

35. As Alfred Marshall (1925) put it: "The richer a man becomes the less is the marginal utility of money to him; every increase in his resources increases the price which he is willing to pay for any given benefit. And in the same way every diminution of his resources increases the marginal utility of money to him, and diminishes the price that he is willing to pay for any benefit" (p. 81).

36. For an elaborate exposition of this argument see Musgrave and Musgrave (1973), pp. 216–19, and Sadka (1976).

37. For an elaborate exposition of this argument see Becker (1971), pp. 52–53.

38. As quoted in Blum and Kalven (1953), pp. 34, 72.

39. For more about the problems associated with models assuming risk-aversion see Brenner (1983b), chapter 1 and its Appendix.

40. Our arguments imply a progressive tax on wealth. The existing progressive taxation schemes have been based on income. This may be due to the fact that it is easier to measure income (why wealth can only be vaguely measured is explained in Brenner 1983b, chap. 2). It may also be illuminating to recall that inheritance tax, a wealth tax, was generally the first progressive taxation measure introduced in most countries.

41. On the emergence of this and other forms of taxation see Ardant (1976) and Comstock (1929).

42. See Brenner (1983b), chapters 1–3.

43. Progressive taxation on both incomes and inheritance was introduced in Prussia in 1891 and, as Ardant (1976) also points out, developed, in part, in response to the socialist movement. Ontario and Nova Scotia introduced these taxation schemes in 1892, while France did so in 1901. In England the practice of taxation implied progressivity from 1800 on (although formally the inheritance-tax rate was proportional, equal to 10%), but became progressive formally in 1894. However, only in 1906 was progressive taxation imposed on incomes. For more examples see the sources quoted in note 41, and for more on the interpretation given to Bismarck's policies see chapter 1, note 13.

44. There is nothing new in this statement. Renan said that "In all human affairs, it is the origins which deserve study before everything else". Ronald Coase (1937) suggested that economists should explain the emergence of some markets, rather than deal with their existing features.

45. This prediction is also examined in a historical context, and with respect to the behavior of groups that have been discriminated against, in Brenner (1983b).

46. What do other individuals do? They gamble on *other* people's ideas (those of their ancestors in particular) and may be defined as followers.

47. This issue is discussed in detail in Brenner (1983b).

48. For more on this point, too, see Brenner (1983b).

49. For a summary of the view that ideas can be perceived as "lucky hits," see Brenner (1983b).

50. Starting from a different set of assumptions, Lazear and Rosen (1981) have shown that such a hierarchy is "efficient."

51. Mary Cunningham, a successful vice-president of the Bendix Corporation, was accused, by gossip, of having secured her promotion in the corporation through her friendship with the president of Bendix, Mr. Agee. She was forced to resign. Later she and Mr. Agee were married. Recently, in a major takeover fight, they were subject to additional gossip; on gossip, in general see Brenner (1983b), chapter 1.

52. The reaction to sudden loss of wealth in my model is ambivalent: one may become either an entrepreneur or a criminal. In the article, "The Chief Executive under Stress," (*New York Times*, 7 November 1982), Barrie Greiff, a psychiatrist at the Harvard Business School, remarked that "Some people really come to the fore when things go wrong. They're the money players." In the same article, Professor Abraham Zaleznik also of the Harvard Business School, said, "I think if we want to understand the entrepreneur we should look at the juvenile delinquent. I think there are a lot of similarities." In addition to John De Lorean, the ex-chairman of General Motors accused of drug dealing, the article mentions Anthony Conrad, the RCA

chairman who did not pay taxes for five years, Eli Black, the United Brands chairman who leaped to his death, leaving behind questions of bribery and impropriety, and others.

53. For more on Schumpeter's views see appendix 1.2.

54. This model shows why employers and employees can bargain. Employees may realize that they have good management and that their productivity elsewhere may be lower: suppose they could only get a $10 hourly wage elsewhere. Suppose that in their present employment they are earning $12 per hour, but they also know that the company's profits are rising. Since it is impossible to determine how much of the increased profits have been due to the better management and how much to the workers' greater efforts, there is place for negotiation.

55. See Drucker's article, "The Innovative Company" (*Wall Street Journal*, 26 February 1982).

56. How insurances based on customs are related to one's wealth and put constraints on one's creativity is explained in Brenner (1983b). The arguments presented there can be illustrated by some recent opinions on the behavior of Italian-Americans. Andrew Rolle, in *The Italian Americans: Troubled Roots*, notes that second-generation Italians grew up with a strong sense of familial obligation and were expected to suffer shame and guilt if they did not follow their fathers' ways—an attitude that leads to "too much" respect for authority. A New York psychologist, Joseph Giordano, director of the American Jewish Committee's Louis Caplan Center on Group Identity and Mental Health, notes that Italian families provide their children with a mixed message: "Make it. Be successful. But don't go too far." This, and other sources, are quoted in Stephen Hall's (1983) article in which he further notes that among Italian-Americans a decline in the family as a traditional symbol of achievement is as likely to provoke crisis as celebration. The model of human behavior that predicts such attitudes is presented in Brenner (1983b).

57. The examination of the Canadian data is based on a research report prepared by Maxime Trottier, Department of Economics, University of Montreal.

58. The term "productivity" (which has never been precisely defined) is used, at times, very strangely by some economists. John Kendrick, for example, who is considered an expert on productivity, wrote in a recent *Fortune* article ("The Coming Rebound in Productivity," 28 June 1982) that "productivity always picks up during recoveries" (can there be recoveries if "productivity" does not pick up?), and that "managements become more cost conscious as profits shrink" (does that mean that they pay less attention to costs when profits increase? How is this view consistent with the economists' assumption of profit maximization and efficiency?). When surveying the economic literature on productivity, it is somewhat frustrating to learn that few writers speak about the link between productivity and either good ideas, the motivation to make greater efforts, or human behavior in general (Leibenstein 1957, though vague, is one exception). Most writers seem so buried in techniques that they forget that their subject matter—human behavior—has been left out. The techniques may be beautiful, elegant, and fascinating intellectual puzzles, but they are only form and gait.

59. For a summary see Klein (1977) and studies in Baldwin and Richardson (1974).

60. For a brief review of the literature see Brenner (1983b), chapter 1. It should be noted that there are some economists who have tried to treat innovations as

endogenous variables—their views are briefly reviewed in Brenner (1983b), chapter 1. See also Scherer (1980).

61. The reason is that the distinction one can make within that approach is between physical and human capital. See Schultz (1961) and Becker (1975).

62. Elon Kohlberg of the Harvard Business School seems to have a perception similar to the one implied by the model: "It has always struck me that Americans are very lonely, particularly the business people. You have to look out for yourselves and be more pragmatic, because you do not have the support systems of other countries—the family and friends you would know for all of your life" (as quoted in Solman and Friedman 1982, p. 72). The views expressed here are formalized and examined in detail in Brenner (1983b), chapters 2 and 3, where the role of the family as insurer is examined. As to another implication of the discussion in this section, notice that it may not be accidental that taxes imposed on imports are called "customs": they are imposed to protect them.

63. Among them: Becker (1957), Leibenstein (1957), Simon (1959), Alchian and Kessel (1962), Marris (1964), Williamson (1964), and Baumol (1977). It should also be noted that many economists have pointed out that omitting the entrepreneur from the theory of the firm may have serious consequences: among them Baumol (1968) and, most prominently, economists belonging to the Austrian school (although they have not provided a workable alternative).

64. Notice later in this section the views of executives on this point: they contradict those of Simon and seem consistent with mine. See also note 17.

65. It is important to emphasize that the term "efficiency" is not defined within my model, since changes in the distribution of wealth are assumed to play the central role when explaining attitudes toward risks, and "wealth" depends on the ideas that people have gambled on. Thus within this model "stability" is the central term, and models built around the assumption that people maximize the utility of their "wealth," which use the word "efficient" to represent some of the outcomes of such models, represent an idea on which some economists gambled.

66. See Brenner (1983b), chapter 1. Notice also that in my model (as in Schumpeter's arguments) "marginal" changes play no role. Only significant gains or losses change people's attitudes toward risks.

67. My views also bear some relationship with Leon Festinger's (1951) "social comparison theory" in the marketing literature. His view was that people evaluate their performance by comparing themselves to others.

68. This is how some executives define the entrepreneurs: " . . . a group of Boeing executives . . . [made the point] that excellent companies seem to take all sorts of special troubles to foster, nourish, and care for what we call 'product champions'—those individuals who believe so strongly in their ideas that they take it on themselves to damn the bureaucracy and maneuver their projects through the system and out to the customer. Someone piped up: 'Champions! Our problem is we can't kill them!' " (Peters and Waterman, 1982, p. xviii).

69. Last, but not least, it should be pointed out that the approach to international trade suggested here (i.e., starting the analysis at the level of firms) is not as unusual as it may seem at first sight: Klein (1977) relied on a similar approach although since he lacks a model, his discussion frequently goes astray. As for the theory of the firm: the points raised in this appendix are elaborated in Brenner (1985a).

Chapter 3

1. See Brenner (1983b). Also notice Schumpeter's (1927) observation that the shift of classes "was the result of outside influence, which was *accidental* from the viewpoint of the class system in existence before" (p. 178; italics added). Fluctuations in population have been linked in Brenner (1983b), chapters 2–4, not only with sudden alterations in the distribution of wealth and the subsequent gambles on ideas, but also with the emergence of new institutions substituting for lost trust.

2. For more on this ambivalence see Brenner (1983b), chapters 1 and 5.

3. For an elaborate treatment of these issues, see Brenner (1983b), chapters 2 and 3.

4. Rudé (1972) also notes that "with industrialization trade unions of hatters, tailors, woolcombers (though illegal), existed already at the beginning of the century, strikes were conducted by tailors from 1719, and many industrial disputes involving sailors and weavers occurred in 1760, which took the forms of both riots and strikes" (p. 285).

5. Some writers, however, emphasize their ambivalent feelings toward this period. Plumb (1972), for example, wrote: "We often think of the middle years of the eighteenth century like creating the dark satanic mills and the desperate conditions of slum life, but those self-same decades also brought the possibility of cultural enjoyment to the mass of mankind" (p. 48). Earlier John and Barbara Hammond's works (1911, 1925) made similar points.

6. For a summary of traditional viewpoints see the articles in the books edited by Cipolla (1973) and Floud and McCloskey (1981), as well as the article by Mokyr (1977) criticizing two of the traditional explanations for the Industrial Revolution (namely, whether or not it was demand or supply induced). McCloskey (1981)—rightly—criticizes them all.

7. Those acquainted with the works of poets, painters, and philosophers? Maybe. But how many people are acquainted with their works and how many just buy cheap clothing (cheap because of innovations made in the past), and therefore will thus know only about the innovations made in the past?

8. In contrast in the previous chapters, when I summarized other people's viewpoints, I did so because there was something in common between their views and mine.

9. The argument that the outburst of inventive activity in England during the Industrial Revolution can be attributed to the fact that this was the first century in which the patent system actually functioned well is not convincing. For two reasons: first, as Nef's works suggest, it may not be true that previous centuries in England did not see such outbursts. Second, one cannot neglect the evidence that the patent system functioned marvelously in primitive societies. Songs, family names, methods of advertising were patented and rights on them were protected by customs (for summary and references see Brenner (1983b), chapter 2. So it is simply not true to say that the eighteenth century was the first one in which the patent system functioned well. However, as explained in Brenner (1983b), when population fluctuates and customs are destroyed, laws, police, and government replace customs in enforcing property rights. England may have been, *by chance*, the first country to make the successful transition. As to creativity: in an additional context Bloch's observation (in his unfinished manuscript published in 1953) should be noted: "It is an established fact that from the twelfth century until at least the Reformation the communities of

textile workers were one of the favorite breeding grounds of heresies" (p. 153). As indicated in the text, textile workers seemed later to be disproportionately represented in riots, and we also know that many innovations were made in this domain. Was this industry subject to more fluctuations than others (in part because there were many innovations, in part because of fluctuating populations which had an effect on perceived benefits of innovating within this domain)?

10. But notice that according to the model and the arguments presented in the appendixes to chapter 1, one cannot distinguish between the terms "divine revelation" and "warmed and overweening Brain."

11. As quoted in Hughes (1971), p. 81.

12. For an elaborate discussion see Hughes (1971), Rosenblum (1971), and their sources.

Chapter 4

1. Before that English inheritance customs were very restrictive. For instance, during the reign of Henri II (1133–89), a man's movable estate was divided into three equal parts: one-third went to his children, one-third to his wife, and the last third he could dispose of as he saw fit (as reported in McCulloch 1848, p. 8). With regard to one's real estate, the law was even stricter; after the Conquest, land could not be legally willed away, although ways were found to do just that (see Pollock and Maitland 1898, 2: 328 ff.). For a historical account of the evolution of inheritance laws in England, see Wedgwood (1929), pp. 20 ff., and Sheehan (1963).

2. Le Roy Ladurie (1976), p. 54.

3. Yver (1966), p. 58.

4. McEvedy and Jones (1978) write: "The Roman Empire prospered until A.D. 200 by which time it had some 46 m subjects, including 28 m of people in Europe. This was . . . followed by a slump which got steadily worse over the next four centuries . . . The decline was general . . . The new trend had dramatic results. The Roman Empire declined and fell . . . and . . . a new society began to form, the feudal society" (pp. 21–22).

5. The direction of the bias is unclear: on the one hand the peerage must have had a longer life expectancy for nonviolent death than the poorer people. This may have meant fewer remarriages. On the other hand, there were more deaths by violence among the peerage than among a group less involved in military acts and duels. Since if the spouse died, the property that had to be protected was greater than for the average man, the nobility had greater incentives to remarry and form new alliances.

6. The Cinderella tale is very old. According to the Opies (1979), the first known Cinderella story was Chinese, dating from A.D. 850. This must not disturb us, since shorter lives and more frequent remarriages may have also characterized ancient Chinese society.

7. We must note that in England there was one kind of tenure, the burgage tenure practiced in towns where by custom the youngest instead of the eldest son inherited. Blackstone (1766, 2: 82–84) says that this tenure antedated the Norman conquest and attributes its existence to a distant pastoral past when all sons but the youngest left their father's home to tend the herds. But he also admits that burgage tenure was insignificant relative to other forms of tenure. Homans (1941) attributed this to the

fact that as the sons who did not inherit were compelled to seek their luck elsewhere, the elder, more mature son who had the best chance of success was the one leaving.

8. In France, even in the provinces where customs enforcing "perfect equality" between heirs existed, the nobles used a male primogeniture principle. See Martin (1972), p. 30.

9. Another example of what the poor noble son could do is provided by McFarlane (1978), quoting an Italian visitor to England in the 1500s: "He [tells] a story of how the brother of the Duke of Suffolk, poor but noble, married the rich old woman with whom he was boarded" (p. 175). (See Schumpeter (1927) on a theory of marriage as a symbol of social class.) We must note that in his book on the difference in the nature of property rights in England and in the rest of Western Europe, McFarlane does not deal explicitly with the reasons for changes in inheritance laws.

10. Also see Brenner (1983b) who argued that many features of the Middle Ages can be viewed as an adaptation to an unexpectedly decreased population with an already achieved military technology.

11. Greenfield, Stricken, and Aubey's (1979) observation should also be noted: "From the beginnings of Portugal's expansion, civil servants and colonial administrators had been drawn primarily from the ranks of the lesser nobility and from among the second and third sons of the wealthy and prestigious upper nobility who did not share in the estates of their families" (p. 24). Such practices, according to the view of human nature presented here, diminish the probability of criminal and rebellious acts taking place.

12. Since ducal families had better health than the rest of the population, this is an upper limit.

13. For a formal proof of these statements see Brenner (1983b), appendix 2.1.

14. This was done by giving the title to the land to a trustee during the lifetime of the owner on condition that the use of the property and its revenues be given to the person designated by the owner (see Holdsworth, 1903, 4: 409, 422 ff.).

15. Land that was not devisable by will was land entailed (see n. 17), held only for a lifetime, or held under another condition. See Blackstone (1766, vol. 2).

16. Apparently, this restriction was imposed for fiscal reasons. If the tenant in knight service died while his heir was still a minor the king was entitled to the revenues of his estate until his majority (the privilege of wardship) and could also sell the right of marrying the heir(ess) (privilege of marriage). These two privileges were an important source of revenue for the king. If this tenant had been allowed to will all his land, the king would have lost both rights. For this reason tenants in knight service were denied complete freedom of will (see Holdsworth, 1903, 4: 450–66). See also Glasson (1882).

17. An entail was a device by which the founder of a family could leave property to his son, grandson, great-grandsons, etc. This right prevented direct descendants from disinheriting their own children.

18. The measure of the devastation brought on the civilian population was such that Bloch (1931) writes that after the Hundred Years' War in French rural areas "you could hear crow neither rooster nor hen" (p. 118). Moreover, Robbins (1928) showed that war rather than the plague explains the characteristic features of French society during this period, despite the fact that it includes the Black Death. This information suggests how devastating the war might have been if a calamity such as the Black Death had not left a greater imprint on society.

19. When comparing French and English development, we must note that even the War of the Roses which struck England at the end of the Middle Ages did not have the same impact on the civilian population (see *Encyclopaedia Brittanica* 15th ed., s.v. "English History"). Moreover, the Renaissance brought conflict to France again, especially with the religious wars (see Heillener 1967).

20. The other question which may be asked is why restrictive inheritance customs appeared in the precise guise they did. Maybe other rules would have served those circumstances as well or even better. Hayek (1973) has emphasized that in any society, knowledge is so fragmented that it is impossible to ascertain which law is the best in the sense of helping society achieve a goal. In consequence, laws and customs emerge, in a sense, randomly. According to Hayek, the laws that persist are those which experience shows to have *survival* value. This is what may have happened with inheritance laws both at the beginning of the Middle Ages and when they were revised, the altered *survival* value being linked to the drastically changed demographic environment. Also see Brenner (1983b), who emphasizes how chance determines the form laws take.

Chapter 5

1. See Brunner (1975) for a discussion of these two issues.

2. See, for example, Lucas (1973) and Fischer (1977).

3. See Gordon's (1975) presentation of a variation on the traditional "cost-push" argument and Brunner's (1975) criticism of it; Brunner criticizes the "cost-push" arguments in general.

4. See Stevens (1981), Feldstein (1981), and the following chapter.

5. See Cagan (1956), Friedman (1969), and Keynes (1923).

6. In part, the arguments in this section draw on Brenner (1979), but at the time that article was written my ideas were vague, and the article was thus less precise. But the article benefited at that time from comments by Donald Moggridge and George Stigler.

7. This is clear from other statements in the *General Theory*: "A fall in real wages due to a rise in prices with money wages unaltered, does not, as a rule, cause the supply of available labour on offer at the current wage to fall below the amount actually employed prior to the rise or prices" (1936, pp. 12–13). Moreover, the definition of full employment in the Appendix to chapter 19 emphasizes this point: " . . . Pigou's book is written on the assumption *that any rise in the cost of living, however moderate relative to the money wage will cause the withdrawal from the labour market of a number of workers greater than that of all the existing unemployed*" (p. 277; italics in original).

8. Although both Dunlop (1938) and Tarshis (1939) have pointed out that this was not the empirical evidence, even in 1939 Keynes defended his view that "money wages are stickier than prices" (p. 36), and he quoted Marshall in his defense, attributing to him the view that "during a slow and gradual fall of prices a powerful friction tends to prevent money wages from falling as fast as prices." Thus at that time Keynes still held to the main structure of the argument (of the first chapter of the *General Theory*) and believed that it needed to be "amended and not discarded."

9. See also Trevithick (1975) and Brenner (1979a). If the idea of "money illusion" was derived from Keynes's other works (see, for example, Maital 1972), it is also hard to understand why. Trevithick's conclusion on this point is the closest to the

view presented here, i.e., that in Keynes's theories inflation serves to provide a politically feasible solution: "Keynes, unlike many neo-Keynesians, did not invoke the assumption of money illusion as an adjunct to his theory of inflation. The inflations which preoccupied Keynes's attention were normally associated with particular situations of national crisis. Keynes would have been the first to admit that his pamphlet ("How to Pay . . . ") should not be regarded as incorporating a general theory of inflation" (pp. 112–13).

10. See, for example, the correspondence with Harrod, who wrote to Keynes that this was not "empirical evidence": "Industrial struggles are clearly not *consciously* struggles about the sharing between particular trades of a given average wage. Every labour man would repudiate this violently. The industrial struggles are almost always intended to strike capitalists, not fellow workers" (Keynes, 1973a, p. 528; italics in original). But Keynes never changed his mind. On the contrary, the word "relative" is emphasized in the following quotation: " . . . the struggle for money wages is . . . , essentially, a struggle to maintain a high *relative* wage . . . " (1936, p. 252).

11. More than a simple notion of "fairness" is hidden behind this question. Its content may be connected with Keynes's view that the demand for labor fluctuates because of the "vagaries of the marginal efficiency of capital as determined by private judgement of individuals ignorant or speculative" (1936, p. 324). Here Keynes implicitly assumes that the distribution of incomes can be based on some "just" basic structure (see chap. 19) which in his model is shaken by changes in expectations that are exogenous to the system and not "justified."

12. Here Keynes implicitly assumes that money wages stay unchanged. We shall return to this point later in the section.

13. Indeed, when Keynes summarizes his policy implications in chapter 19 of the *General Theory*, his motto is: "only a 'foolish' or 'unjust' or 'inexperienced' person would prefer a flexible wage policy to a flexible money policy" (p. 268–69).

14. "Temporarily sinking" in the sense of decreased wealth.

15. See especially the correspondence with Hawtrey (especially 1973b, pp. 18, 24, 32), or the correspondence with Robertson and Harrod (especially 1973a, pp. 501–2, 528–34). Only Joan Robinson seemed to understand Keynes's intentions: "I found the chapters on . . . money wages particularly well done, but all of it is as clear as this sort of stuff can be" (1973a, p. 638), whatever this clear sentence means.

16. "In regard to his [Viner's] criticism of my . . . treatment of involuntary unemployment, I am ready to agree that this part of my book is particularly open to criticism . . . and [I hope] to make improvements" (1973b, p. 110).

17. That animal spirits are exogenous to the economic system, in Keynes's view, is well known. As he pointed out in his correspondence: "The theory [of the *General Theory*] can be summed up by saying that given the psychology of the public, the level of output and unemployment as a whole depends on the amount of investment. [But the factors] which determine the rate of investment . . . are most unreliable, since it is they which are influenced by our views of the future" (1973b, p. 12). See also note 11: in the quotation used there Keynes attributes the change in expectations to the "vagaries of the marginal efficiency of capital" determined by ignorants.

18. This inflationary process refers to the *General Theory* only. In Keynes's other works, like the *Tract*, or "How to Pay for the War," the inflationary processes

described are different. See Trevithick (1975). Before turning to additional statements in the *General Theory*, and to other works by Keynes, which show that in his
arguments Keynes relied many times on some vague notions of justice, let us first
briefly mention other explanations given to the quotations used in building the
arguments. One such explanation was that the worker's "views of what his services
should be worth unavoidedly are related to what he was paid only yesterday. His
expectations are 'inelastic' in the Hicksian sense . . . " (Leijonhufvud 1968, p. 96).
In this case the worker uses either his previous wage, or the wages of others, as a
piece of information in determining his reservation wage. Although this argument is
consistent with modern search theories, it contradicts Keynes's argument, as no
asymmetrical behavior with respect to the source of the decrease in real wages can be
obtained. One may thus conclude that the aforementioned quotations in the *General
Theory* are only to be used for defining the "involuntarily unemployed" and
Keynes's notion of "justice."

19. See also Brenner (1980). In several other places throughout the *General
Theory* we can find statements showing that Keynes indeed assumed the existence of
nominal wage contracts: "Thus we can sometimes regard our ultimate independent
variables as consisting of . . . the wage unit as determined by the bargains reached
between employers and employed" (1936, pp. 246–47), or "A movement by employers to revise money wage bargains downward will be much more strongly
resisted than a gradual and automatic lowering of real wages as a result of rising
prices" (p. 264). Of course, unless there were nominal contracts to be revised, this
sentence is meaningless.

20. This was also Lerner's (1952) conclusion on the role of nominal contracts in
Keynes's theory: "The usefulness of money depends intimately on a certain degree
of stability in its purchasing power. This stability encourages the making of contracts
and the development of other institutions that help to establish a rigidity of the
general price level. The rigidity in turn gives further stability to the purchasing power
of money so that its position in the economy reinforces itself in the mould of custom"
(p. 192).

21. It is because of the existence of such contracts that Keynes also criticized the
whole concept of "neutral money" and stated that " . . . booms and depressions are
phenomena peculiar to an economy in which money is not neutral" (1973a, p. 411).
In the same paper Keynes wrote: "Pigou knows as well as anyone that wages are in
fact sticky in terms of money. Marshall was perfectly aware that the existence of
debts gives a high degree of practical importance to changes in the value of money"
(p. 410). It is clear from this quotation that Keynes refers to monetary debts, and
that the words "sticky wages" refer to nominal wage contracts. And later Keynes
stated that "neutral money" was a "nonsense notion" (1973b, p. 93).

22. This part of the section draws on Brenner and Mishkin (1978).

23. There are many additional quotations that show the similarity between
Fisher's and Keynes's arguments: "The harm arises from the fact that there are in the
conduct of business other money transactions besides the purchase and sale of
commodities and that those other transactions do not share in the effects of inflation,
deflation . . . If they did, inflation and deflation would do absolutely no harm.
Everything would move to scale" (Fisher, 1934a, p. 35; italics in original).

24. This view is reflected in the following two quotations:

The bondholder is . . . the "silent partner" in business. He lacks . . . the training to be a risk bearer or captain of industry. But after years of falling price level, . . . the captain [of industry] is held responsible for the shipwreck, is forced out, discredited, [although] the fault has arisen from this unreliable instrument of reckoning, the dollar. Next, the bondholders or their representatives, often lawyers, take control whether or not they know how to run the business. [Fisher 1928, pp. 104–5]

It might [be argued] that since the debtors gain exactly as much as the creditor loses no harm can be done to society as a whole either by inflation or deflation, since the average wealth would not be changed . . . But the losses exceed the gains owing to the indirect harm of uncertainty, unemployment, strikes . . . These can only mean a dead loss to the general public. [pp. 102–3]

More about these arguments in the next section of the text.

25. See the correspondence between Keynes and Hicks, pp. 71–77, in *The General Theory and After II* (1973b).

26. Hicks never defines "unemployment"; he says only "that if there is a great deal of unemployment" and if wages are unchanged, then Keynes's model holds.

27. Hicks had another reservation, and it is interesting to note that he did not connect the problem of money wages with it: " . . . one cannot escape the impression that there may be other conditions when expectations are tinder, when a slight inflationary tendency can light them up very easily . . . then an increase in [money] income will raise the . . . rate of interest . . . " (p. 158).

28. See Leijonhufvud (1968) and Robinson (1962).

29. For the formal proof see appendix 1.2.

30. The subject of expenditures on armament will be discussed later in further detail.

31. Mandeville (1932) repeats and clarifies his viewpoint in these quotations:

But would you have a frugal and honest Society, the best Policy is to preserve Men in their Native Simplicity, strive not to increase their Numbers; let them never be acquainted with Strangers or Superfluities, but remove and keep from them every thing that might raise their Desires, or improve their Understanding. [pp. 184–85]

The Great Art then to make a Nation happy and what we call flourishing, consists in giving every Body an Opportunity of being employ'd; which to compass let a Government's first care be to promote as great a variety of Manufactures, Arts, and Handicrafts, as Human Wit can invent; and the second to encourage Agriculture and Fishery in all their Branches, that the whole Earth may be forc'd to exert itself as well as Man; for as the one is an infallible Maxim to draw vast Multitudes of People into/a Nation, so the other is the only Method to maintain them. [p. 197]

32. How the subject of wars can be dealt with within the model summarized here is discussed in chapter 1.

33. One can immediately ask how this view can be tested. For one can argue that the researchers for NASA and the Pentagon gave biased interpretations (in line with their self-interest, as Stigler 1975 suggests); knowing that the expenditures are not "profitable" they attribute a nonpecuniary benefit to their consequences. If that were the case for the space program and, let's say, for a large part of the nation's

military expenditures, diminished wealth could be expected to amount to tens of billions of dollars. As the most simplistic quantity theory would predict, such perceptions would result in a certain inflation rate. However, if the perception was that the expenditures are a necessary insurance, there would be no inflation. So the viewpoint is falsifiable.

34. It may be useful to elaborate slightly the relationship between expenditures on arms, wealth, and the viewpoints of Keynes and Mandeville. In general, one can say that when, for some reason, aspirations start to be lowered, the probability that people will bet on new ideas—either "entrepreneurial" or "political"—increases. But, a priori, one cannot say in what sphere an individual's new idea will catch fire: will it be a new innovation in technology, business, or science, or will it be a political one that restores confidence? That depends on both historical circumstances and the emergence of "the right man at the right time" (for example, David inventing a new weapon). But the emergence of such a man is a matter of chance. Consider the following scenario: A country is outdone by another and a sputnik is sent to the skies. A political leader may emerge who promises to send one there quickly and redress the balance. The idea may catch fire in the minds of men: the government will spend money on a new space program. If people's perception is that such activities indeed are likely to restore their wealth, they will exert themselves more and the increased government expenditures may not necessarily crowd out private investment. Notice, however, that the increased economic activity is no *"free* lunch"—such reactions take their emotional toll. (But this, of course, is hardly measurable.)

The parallel with Mandeville, Keynes, and the mercantilists, with whose views Keynes sometimes agreed, is obvious. Keynes (1936) summarized this view as follows: "The mercantilists were under no illusion as to the nationalistic character of their policies and their tendency to promote war. It was *national* advantage and *relative* strength at which they were admittedly aiming" [p. 348, italics in original]). Were nations suddenly to trust one another, could one conclude that amounts previously spent on weapons would suddenly be all invested in the private sector? No model of human behavior makes such a prediction. The answer is, in fact, negative (as Mandeville and Keynes have suggested): once threats diminish, people exert themselves less. As *measured* today, the GNP would as a consequence diminish because of the greater trust (also, unemployment would suddenly rise and the distribution of wealth would change—but the latter two effects are costs societies must always pay for transitions from one order to another).

35. This conclusion should not be surprising to anybody who is familiar with what is called "the scientific method," by which "scientific" ideas and their predictions are tested by a "stop and go" policy. In the context of this study political ideas are being examined by a similar "methodology." Why should the methods for testing ideas differ?

36. See Ian McKinnon (1983).

37. Other mistakes, like tying social-security benefits and family allowances to inappropriate indexes, and their effects are discussed in the next chapter.

38. Interestingly, Friedman (1984) recently made a similar point, without noticing that such an approach provides a rationale for a discretionary monetary policy, if one wants to maintain a stable price level.

39. At the risk of repeating myself, let me put this framework issue more bluntly: it has been shown in this chapter that Keynes's view of human behavior may have

been radically different from the one assumed by the neoclassical approach. It is this difference that lies at the root of some of his policy recommendations, in particular that of the role of government in getting an economy out of crisis, and seems to shed light on his attitudes toward government deficits, savings, and "crowding out." It follows that the debate among believers in IS-LM–type models and monetarists (who assume "stationary" processes) may be rather irrelevant—the basic question is: which view of human behavior is accurate, the one suggested here or the neoclassical one? If the interpretation of some of Keynes's views given here is accurate, the discussion on macroeconomic policies, on crowding out in particular, should be approached from a radically different angle. Thus the models of Barro (1974) and Blinder and Solow (1973), those summarized in Carlson and Spencer (1976), the discussions of Modigliani (1983), Benjamin Friedman (1983), Blanchard (1983), or those found in the volume *Deficits* edited by Conklin and Courchene (1983) (with the exception of McKinnon's article), cannot shed much light on the subject. But the approach presented here is in line with some of the views of Friedman (1983) who emphasizes that the problem of deficits is political rather than economic.

40. See Fisher (1932, 1933a, b).

41. However, it is interesting to note that while, according to "rational expectations" models, individuals are "cheated" and are induced to behave nonoptimally when they supply a greater amount of labor, according to Keynes the individuals behave in such a way because they are "fair." Thus, while in the "rational expectations" models individuals should be "unhappy" (having a decreased utility) when they discover that they were cheated, in the *General Theory* they agree to the decrease in real wages "happily."

42. Not only recently, but also during the periods of hyperinflation.

43. See Fisher (1911, 1928, 1932, 1933a, 1934a, b): Fisher repeats his views in all of these references.

44. For a definition of the "NRU" see Phelps (1970).

45. See Brenner (1985b), Brenner and Courville (1985).

Chapter 6

1. This section is based on the first chapter of my Ph.D. thesis, published in *History of Political Economy* (see Brenner 1979b). At the time of its writing I benefited from comments by Nissan Liviatan, Joel Mokyr, Frederic Mishkin, and Don Patinkin.

2. See Schumpeter (1954), p. 1091.

3. See Schumpeter (1954), p. 528.

4. If there was indexation it would eliminate the option of governments to redistribute wealth through inflation.

5. Early drafts of this section were prepared in collaboration with, and with financial support from, the C. D. Howe Institute in Toronto. The views presented here do not necessarily reflect the views of the institute.

6. For background information on indexing in the nontechnical literature see Friedman (1974a, b, c), Lacroix and Montmarquette (1975), and Brenner (1979b).

7. This point has been made by Jo Anna Gray (1976), Liviatan and Levhari (1977), Brenner (1978), and Cousineau and Lacroix (1981). But see note 19 on why indexed contracts may sometimes have shorter duration than nominal ones.

8. For a detailed description of the experiments with indexation in these countries see Brenner (1978).

9. However, see note 17.

10. The benefit of indexing wages is not to raise the value of indexed contracts above that of nonindexed ones, since either the latter can be renegotiated when their value suddenly falls, or employers may implicitly commit themselves to compensate in the forthcoming negotiation for the temporary loss when inflation rates unexpectedly climb. I owe this interpretation to Robert Lacroix, who, in a study written with F. Dussault (1981), shows that the data do not contradict it.

11. For an elaborate discussion on the various indexation arrangements in Canada, see Wilton (1980); for arrangements in other countries see Brenner and Patinkin (1977) and Brenner (1978).

12. See *Canada Year Book 1976–77* (1977), pp. 998–99.

13. These seem to be the implicit arguments behind the process called "cost-push" inflation. See Gordon (1975) and comments on Gordon's paper by Brunner (1975). See also Laidler and Parkin (1975) and Frisch (1977). But see previous chapter on criticism of the standard "cost-push" analysis.

14. See also Brenner and Patinkin (1977).

15. Although notice latter reservations when price indexes are discussed.

16. See *Canada Year Book 1976–77*, p. 275.

17. On some difficulties involved in computing price indexes, see Wallace and Cullison (1979).

18. See Blinder (1980).

19. The increased uncertainty as to the meaning of changes in the CPI during inflationary periods is discussed in Parks (1975), Blinder (1980), and Stevens (1981) and may provide an explanation as to why at times indexed contracts may be of shorter duration than those determined in money terms.

20. See Brenner and Patinkin (1977) and Brenner (1978).

21. See Stevens (1981) and Feldstein (1981).

22. See Brenner (1978).

23. See *Canada Year Book 1976–77*, p. 275.

24. This summary is based on Brenner (1978).

25. See Brenner and Patinkin (1977).

26. Although it should be noted that indexation has been used in a way that has provided greater protection for lower wages. The indexation clause that has achieved this goal has been the following: the most typical formula for a COLA of wages relates an absolute movement in the CPI to an absolute change in the wage rate. This means that a one-point increase in the CPI might be worth a 2.5¢ per-hour wage increase to all workers within a bargaining unit. Since the same adjustment is given to all workers regardless of their wage rate, these adjustments provide relatively less protection for workers who are paid more than the average rate. In a sample of 328 indexed contracts, 94.4% were based on these types of adjustments and only 5.6% were based on percentage-point adjustments which provide equal protection for all wages (see Wilton 1980). This greater automatic protection given to lower wages is also typical of indexed contracts in other countries, only their method of achieving it is more explicit. In Belgium, salaries above $1,000 a month were frozen in 1974. In Israel incomes below a ceiling were compensated for 80% of the change in the CPI in

some years, 70% in others, whereas incomes above the ceiling were compensated with a fixed nominal sum equal to the maximum compensation given to the ceiling income (see Brenner and Patinkin 1977).

27. Note should be taken of an event that happened in Canada and not elsewhere. In 1975 the Anti Inflation Board was set up to control wages and profit margins. The legislated wage guidelines with respect to changes in the CPI were as follows: wage increases were allowed to recover the full increase in the CPI with a one-year lag. If employees were to expect such guidelines to be enforced, full protection against all changes in the CPI would be implied, and no formal indexation would be needed. The diminished percentage of formally indexed contracts after 1975 (see table 6.3) may thus be attributed to the fact that the legislated guidelines substituted for them. But this interpretation of the data does not seem very convincing since several facts cannot be explained by relying on it. For example: why then, were 38% of the employees still covered by formally indexed contracts in 1976? Why does one observe both before and after 1975 several types of partially indexed wage contracts? The interpretation that can be given to the diminished percentage of formally indexed contracts after 1975 is the same as the one given to the fact that in 1974 indexed contracts had a shorter duration than nonindexed ones: namely, the greater uncertainty surrounding the meaning of changes in price indexes after 1975. This greater uncertainty after 1975 could be caused not only because of the continuing high inflation rates and the resulting distortions in the CPI, but also because of the 1975 policies of price and wage controls. During and after controls, price indexes lose much of their economic content. In conclusion, the guidelines may have played no roles in wage negotiations: if both employers and employees perceived that they could be harmful and lead to layoffs, they could simply ignore them.

28. For example, Finch (1956), Gennaro (1975), Goode (1951), Knox (1964), etc. However, only recently have more formal models of indexation been developed. See S. Fischer (1975), Liviatan and Levhari (1977), and Brenner (1978).

29. An indirect result of indexing bonds issued by the government would be that, because of competition for funds, the private sector would be obliged to issue indexed bonds as well (see Friedman, 1974b). In practice, however, this did not happen (see Brenner, 1978).

30. See Grossman (1975), Friedman (1974a), and Hicks (1974).

31. Again, Friedman (1974a).

Chapter 7

1. Most recently McNeill (1982), although individuals and their ambitions, and classes and their ideologies do not always seem to play significant roles in the book.

2. See the jealousies described in Johanson and Edey's (1981) *Lucy*, in Martin Gardner's (1981) *Science: Good, Bad and Bogus*, to mention just a few of the works that do not always provide a flattering view of the scientific community ("community"?!). See also the recently published *Betrayal of Truth* by Broad and Wade (1982). This view suggests that Kuhn (1970) might be wrong when he presents the activity called "scientific research" as characterized by a fight among "paradigms." "Paradigms" do not fight—people do, and so it is their perceptions that must be taken into account in order to understand the behavior in this branch of human activity. To put it in more precise terms: a new idea destroys the acquired "human

capital" (what a robust term for such fragile content) and may lead either to entrepreneurial acts or criminal ones (including possibly the forming of a "cartel" to try to prevent the diffusion of the new idea). No wonder that Machiavelli, who avoided all the established categories in the social sciences of his time *and* their vocabulary, was harshly criticized by his contemporaries, that Spinoza was excommunicated, that Rousseau was hounded for half of his life, in 1765 even stoned in Moûtiers, that the universities ignored Schopenhauer and his books "as if to substantiate his claim that all advances in philosophy are made outside academic walls." "Nothing," says Nietzsche, "so offended the German savants as Schopenhauer's unlikeness to them," . . . [and he concluded that] "Experience teaches us that nothing stands so much in the way of developing great philosophers as the custom of supporting bad ones in state universities . . . But [Schopenhauer] had learned some patience; he was confident that, however belated, recognition would come. And at last, slowly, it came. Men of the middle classes—lawyers, physicians, merchants— found in him a philosopher who offered them no mere pretentions jargon of metaphysical unrealities, but an intelligible survey of the phenomena of actual life" (quoted in Durant 1926, pp. 307, 410). Schopenhauer writes boldly, clearly, and— surprise—with a sense of humor. No wonder he was disliked. Not that other innovators fared better with most of their contemporaries: the establishment found Voltaire and his humor too much to take, and Marx's and Veblen's lives were, for a while, quite miserable.

3. How did Barbara Tuchman write? "Historians can—though not all do—make themselves understood in everyday English, the language in use from Chaucer to Churchill. Let us beware of the plight of our colleagues, the behavioral scientists, who by use of a proliferating jargon have painted themselves into a corner—or isolation ward—of unintelligibility. They know what they mean, but no one else does. Psychologists and sociologists are the farthest gone in the disease and probably incurable. Their condition might be pitied if one did not suspect it was deliberate. Their retreat into the arcane is meant to set them apart from the great unlearned, to mark their possession of some unshared, unsharable expertise" (1981, p. 55). But Tuchman does not provide an answer to the question *why* social scientists invented these languages and continue to use them. My model gives some very uncomfortable answers: In fact, it suggests an answer to the question why the revision of old theories in the light of new facts is neither as complete nor as fast as one might expect. Belief in theories does not always depend on long experience: people gamble on an idea after having just one catastrophic observation (Keynesians after the Great Depression, for example). Later they may stick to this idea, which will determine the kind of "facts" they will perceive (in the case of Keynesians, aggregates, and nobody quite knows what they measure).

4. See Schumpeter (1927), pp. 154–69.

5. In several chapters of *History—The Human Gamble* I made a similar point: that "too much" success destroys creativity (in the arts in particular). Here are some additional thought-provoking examples: not only Rossini, but also Sibelius, Ives, Elgar, and Copland gave up composing at the height of their careers. Recently, Bjorn Borg, the tennis "superstar," explained his withdrawal in these words: "There was something missing inside me . . . There is a huge difference between No. 1 and No. 2." Rossini, as quoted in the previous book, said something similar: when one is

on the top, while improvements might be possible, why bother? Who would be able to tell that indeed there was an improvement? So Sibelius drank, Rossini cooked, Ives sold insurance, and Borg advertises.

6. Although he does not use these words. See the discussion on his views in Brenner (1983b).

7. See Wilson (1983).

8. See Berlin (1981), whose words are quoted in appendix 1.1.

9. But it is useful to repeat Berlin's (1981) observation that the word "labor" does not emerge as a clear concept from Marx's work: sometimes Marx speaks of labor as "identical with that free creation which is the fullest expression of untrammelled human nature, the essence of happiness, emancipation, frictionless rational harmony within and between men. At other times he contrasts labour with leisure; and promises that with the abolition of the class war labour will be reduced to a minimum" (p. 95). The difference with my views is clear: what Marx seemed to admire as "the fullest expression of human nature" owes its existence, within my model, to perceived inequality. Moreover, I don't use the undefined word called "labor."

10. There are additional differences from Marx's views when one examines narrower issues: for example, in contrast to views on the "automatism" of accumulation, which says that the rich get richer and the poor poorer, my views suggest that unless a regime is settled (in which the rich stay rich, and the poorer just stay poorer), entrepreneurs who come from among people who became poorer will become richer. (Also, one should note that "captured surplus value does not invest itself"— *somebody* is deciding). See Schumpeter (1927) on this subject, among others, who notes that

> No matter which area we study, we always find that the relative position of families in the class situation undergoes change, not in such a way that the "big" ones grow bigger and the "small" ones smaller, but typically the other way round. In the textile area of Brno, the silk region of Krefeld, the ironworking district around Birmingham, for example, certain families have maintained their position for more than half a century, in many cases considerably longer. Yet, by and large, the families that led around the middle of the nineteenth century are not on top of the heap today. Some of those that are most successful now were then scarcely recognized as members of the class, while some of those that were most successful then are accepted only with reservations today. [p. 154]

11. See Cohen (1977) and Brenner (1983b), chapter 2.

12. See Cohen (1977), Boserup (1965, 1981), Brenner (1983b), and a large number of other writers summarized in Cohen (1977).

13. See Bloch (1940), North and Thomas (1973), Brenner (1983b), chapter 3, and others quoted in chapter 4 of this book.

14. See Brenner (1983b).

15. See appendix 2.2 on international trade.

16. North's (1981) observation should be noted, too: "Population pressure, then as now, is a two-edged sword. It is a major factor in internal and international conflict, political instability, and decline; but here I follow the cutting edge that induced societies to innovate forms of political-economic organization that pro-

moted productivity increase leading to periods of sustained economic growth" (p. 110).

17. See Thomas (1974).

18. This observation is not novel. Sir Henry Maine in his essays on *Popular Government* made it in 1885: "All that has made England famous, and all that has made England wealthy, has been the work of minorities, sometimes very small ones. It seems to me quite certain that, if for four centuries there had been a very widely extended franchise and a very large electoral body in this country, there would have been no reformation of religion, no change of dynasty, no toleration of Dissent, not even an accurate Calendar. The threshing-machine, the power-loom, the spinning-jenny, and possibly the steam-engine, would have been prohibited" (p. 112). The discussion on minorities in general can be found in Brenner (1983b).

19. One may immediately ask how China fits into the relationship between wars and population? The answer is that the populations of China and Japan grew, but did *not* fluctuate significantly. When the population grows slowly, customs can be adjusted and they are not destroyed—for more on this point see Brenner (1983b). (See also appendix 1.3 on political thought.)

20. Also see Billington (1980) who mentions the role of music in the same context.

21. Many other viewpoints which suggest an ambivalent attitude toward "progress" are summarized in Waelder (1967). He notes that in the middle of the sixteenth century Lorenzo Valla called the mariner who sails his ship on an uncharted course or the doctor who tries out new medicine on the sick contemptible. In contrast to this and other opinions expressed there, my approach is different: I try to point out the *choices*, and interpret policies in their light. Whether or not "progress" or "development" (as the term is defined in the last section of chapter 2 of *History—The Human Gamble*) is "good" or "bad" will depend on the reader's viewpoint.

References

Abramovitz, Moses. 1956. Resource and output trends in the United States since 1870. *American Economic Review* 46:5–23.

Alchian, Armen A. 1953. The meaning of utility measurement. *American Economic Review* 42:26–50.

Alchian, Armen A., and Harold Demsetz. 1972. Production, information costs and economic organization. *American Economic Review* 62:777–95.

Alchian, Armen A., and Reuven Kessel. 1962. Competition, monopoly, and the pursuit of pecuniary gain. In *Aspects of labor economics*, ed. H. G. Lewis. Princeton, N.J.: National Bureau of Economic Research.

Allen, David P. 1952. *The nature of gambling*. New York: Coward-McCann.

Angell, Norman. 1933. *The great illusion* New York: Putnam.

Ardant, G. 1976. *Histoire financière de l'antiquité a nos jours*. Paris: Gallimard.

Ariès, Philippe. 1973. *L'enfant et la vie familiale sous l'Ancien Régime*. Paris: Editions du Seuil.

Aristotle. *Politics*. New York: Penguin Books, 1980.

Arrow, Kenneth J. 1970. *Essays in the theory of risk bearing*. Amsterdam: North Holland.

Ashton, John. 1898. *The history of gambling in England*. Montclair, N.J.: Patterson Smith, 1969.

Ayres, C. E. 1944. *The theory of economic progress*. Chapel Hill: University of North Carolina Press.

Baldwin, Robert E., and J. David Richardson. 1974. *International trade and finance*. Boston: Little, Brown and Co.

Barro, Robert J. 1974. Are government bonds net wealth? *Journal of Political Economy* 82:1095–1117.

Bateson, Gregory. 1972. *Steps to an ecology of mind*. New York: Ballantine Books.

Baumol, William J. 1968. Entrepreneurship in economic theory *American Economic Review* 59:64–71.

———. 1977. *Economic theory and operations analysis*. Englewood Cliffs, Prentice-Hall.

Beaton, Albert E. 1975. The influence of education and ability on salary and attitudes. In *Education, income and human behavior*, ed. Thomas F. Juster. New York: McGraw-Hill.

Beattie, J. M. 1972. Towards a study of crime in 18th century England: A note on indictments. In *The triumph of culture: 18th century perspectives*, ed. F. Paul and D. Williams. Toronto: A. M. Hakkert.

———. 1977. Crime and the courts in Surrey 1736–1753. In *Crime in England, 1550–1800*, ed. J. S. Cockburn. Princeton, N.J.: Princeton University Press.

Becker, Gary S. 1957. *The economics of discrimination*. Chicago: University of Chicago Press.

———. 1971. *Economic theory*. New York: Alfred A. Knopf.

———. 1975. *Human capital*. New York: Columbia University Press.

———. 1981. *A treatise on the family*. Cambridge: Harvard University Press.

———. 1983. The fire of truth: A remembrance of law and economics at Chicago, 1932–1970 (discussion, edited by E. W. Kitch). *Journal of Law and Economics* 24:163–234.

Bellamy, Edward. 1887, *Looking Backward 1887–2000*, New York: Random House, 1951.

Benassy, Jean-Pascal. 1975. Neo-Keynesian disequilibrium theory in a monetary economy. *Revue of Economic Studies* 42:503–23.

Bergler, Edmund. 1957. *The psychology of gambling*. New York: Hill and Wang.

Berlin, Isaiah. 1939. *Karl Marx*. 4th ed. Oxford: Oxford University Press. 1978.

———. 1981. *Against the current*. Oxford: Oxford University Press.

Bernard, L. L. 1944. *War and its causes*. New York: Henry Holt & Co.

Billington, James H. 1980. *Fire in the minds of men*. New York: Basic Books.

Blackstone, William. 1766. *Commentaries on the laws of England*. Vol. 2. Oxford: Clarendon Press.

Blanchard, Olivier J. 1983. Current and anticipated deficits, interest rates and economic activity. Discussion Paper no. 998, Harvard Institute of Economic Research.

Blaug, M. 1976. Kuhn versus Lakatos, or paradigms versus research programmes in the history of economics. In *Method and appraisal in economics*, ed. S. J. Latsis. Cambridge: Cambridge University Press.

Blinder, Alan S. 1980. The consumer price index and the measurement of recent inflation. *Brookings Papers on Economic Activity*, 539–65.

Blinder, Alan S., and Robert M. Solow. 1973. Does fiscal policy matter? *Journal of Public Economics* 2:319–37.

Bloch, Marc. 1931. *Les caractères originaux de l'histoire rurale française*. 1931. Paris: Société d'edition "Les belles lettres."

———. 1940. *La société féodale*. 1940. Translated by L. A. Manyon: *Feudal society*. Chicago: University of Chicago Press, 1961.

———. 1953. *The historian's craft*. New York: Alfred A. Knopf.

Blum, W. J., and H. Kalven, Jr. 1953. *The uneasy case for progressive taxation*. Chicago: University of Chicago Press, 1966.

Boissonade, Gustave M. 1873. *Histoire de la réserve héréditaire et de son influence morale et économique*. Paris: Guillaume.

Bolen, Darrell W., and William H. Boyd. 1968. Gambling and the gambler: A review of preliminary findings. *Archives of General Psychiatry* 18:617–30.

Boserup, Ester. 1965. *The conditions of agricultural growth*. Chicago: Aldine.

———. 1981. *Population and technological change*. Chicago: University of Chicago Press.

Brenner, Gabrielle A. 1985a. Why did inheritance laws change? *International Review of Law and Economics*, 5.

———. 1985b. Why do people gamble? Further Canadian evidence. In *The gambling studies: Proceedings of the Sixth National Conference on Gambling and Risk-Taking*, ed. W. R. Eadington. Reno: Bureau of Business and Economic Research, University of Nevada.

Brenner, Gabrielle A., and Reuven Brenner. 1981. Why do people gamble? Rapport de Recherche no. 81–11, Ecole des Hautes Etudes Commerciales.

———. 1982. On memory and markets, or why are you paying 2.99 for a widget? *Journal of Business* 55:147–58.

———. 1983. Qui sont les acheteurs des billets de loterie? *Revue Internationale de Gestion* 8:27–30.

Brenner, Harvey M. 1973. *Mental Illness and the Economy*. Cambridge: Harvard University Press.

Brenner, Reuven. 1978. Micro and macro-economic aspects of indexation. Ph.D. dissertation. Hebrew University.

———. 1979a. Unemployment, justice and Keynes' *General Theory*. *Journal of Political Economy* 87:837–50.

———. 1979b. The concept of indexation and monetary theory. *History of Political Economy* 11:395–405.

———. 1980a. The role of nominal wage contracts in Keynes' *General Theory*. *History of Political Economy* 12:582–87.

———. 1983a. Betting on ideas—Is this the meaning of competition? Report no. 8304. Department of Economics, University of Montreal.

———. 1983b. *History—The Human Gamble*. Chicago: University of Chicago Press.

———. 1983c. Why does the U.S. export new technologies and import older, perfected products?—An alternative view of international trade. Report no. 8320, Department of Economics, University of Montreal.

———. 1984. Crowding-out and the Fable of the Bees. Study prepared for the Economic Council of Canada.

———. 1985a. The theory of the firm—An alternative viewpoint. Report no. 8502, Department of Economics, University of Montreal.

———. 1985b. State-owned enterprises—Practices and viewpoints. Economic Council of Canada.

Brenner, Reuven, and Leon Courville. 1985. Industrial strategies—An alternative viewpoint. Study prepared for the MacDonald Commission.

Brenner, Reuven, and Frederic S. Mishkin. 1978. Keynes and Fisher on the interaction of nominal contracts, changes in the price level and entrepreneurial skill. Revised version of Report no. 7905, Department of Economics, University of Chicago.

Brenner, Reuven, and Don Patinkin. 1977. Indexation in Israel. In *Inflation theory and anti-inflation policy*, ed. E. Lindberg. London: Macmillan.

Bridenbaugh, Carl. 1968. *Vexed and troubled Englishmen 1590–1642*. New York: Oxford University Press.

Broad, William, and Nicholas Wade. 1982. *Betrayers of the truth: Fraud and deceit in the halls of science*. New York: Simon and Schuster.

Brunk, Gregory G. 1981. A test of the Friedman-Savage gambling model. *Quarterly Journal of Economics* 96:341–48.

Brunner, Karl. 1975. Comments. *Journal of Law and Economics* 18:807–36.

Buchanan, James M. 1979. *What should economists do?* Indianapolis: Liberty Press.

Cagan, Phillip. 1956. The monetary dynamics of hyperinflation. In *Studies in the quantity theory of money*, ed. M. Friedman. Chicago: University of Chicago Press.

Canada Year Book 1976–77. Special ed. Ottawa: Statistics Canada.

Carlson, Keith M., and Roger W. Spencer. 1976. Crowding-out and its critics. In *Current issues in monetary theory and policy,* ed. T. M. Havrikesky and J. T. Boorman. Arlington Heights, Ill.: AHM.

Carr-Hill, R. A., and N. H. Stern. 1979. *Crime, the police and criminal statistics.* New York: Academic Press.

Chafetz, Henry. 1960. *Play the devil: A history of gambling in the United States from 1492 to 1955.* New York: Clarkson N. Potter.

Choucri, Nazli, and Robert North. 1974. *Nations in conflict.* San Francisco: W. H. Freeman.

Cipolla, Carlo, ed. *The economic decline of empires.* London: Methuen.

Cipolla, Carlo M. 1973. *The Industrial Revolution.* Glasgow: Fontana.

———. ed. 1976. *Before the Industrial Revolution: European society and economy, 1000–1700.* New York: W. W. Norton.

Clausewitz, Carl von. 1832. *On War,* ed. A. Rapoport. New York: Penguin Books, 1982.

Coase, Ronald H. 1937. The nature of the firm. *Economica* 4, n.s.: 386–405.

———. 1974. The market for goods and the market for ideas. *American Economic Review* 64:384–91.

Cockburn, J. S. 1977a. The nature and the incidence of crime in England 1559–1625: A preliminary survey. In *Crime in England 1550–1800,* ed. J. S. Cockburn. Princeton, N.J.: Princeton University Press.

———. 1971b. *Crime in England 1550–1800.* Princeton, N.J.: Princeton University Press.

Cohen, Mark N. 1977. *Food crisis in prehistory.* New Haven: Yale University Press.

Comstock, A. 1929. *Taxation in the modern state.* New York: Longmans, Green.

Conklin, W. David, and Thomas J. Courchene, eds. 1983. *Deficits: How big and how bad?* Toronto: Ontario Economic Council.

Cooper, J. P. 1976. Patterns of inheritance and settlement by great landowners from the fifteenth to the eighteenth century. In *Family and inheritance,* ed. J. Goody et al. Cambridge: Cambridge University Press.

Cornish, D.B. 1978. *Gambling: A review of the literature and its implications for policy and research.* London: HMSO.

Coulton, G. G. 1926. *The medieval village.* Cambridge: Cambridge University Press.

Cousineau, Jean-Michel, and Robert Lacroix. 1981. *L'indexation des salaires.* Manuscript, Department of Economics, University of Montreal.

Craig, Gordon A. 1978. *Germany 1866–1945.* Oxford: Clarendon Press.

Danziger, Sheldon, and David Wheeler. 1975. The economics of crime: Punishment or income distribution? *Review of Social Economics* 33:113–31.

Darnton, Robert. 1984. *The great cat massacre.* New York: Basic Books.

Deane, Phyllis. 1973. The Industrial Revolution in Great Britain. In *The emergence of industrial societies,* ed. Carlo M. Cipolla. Glasgow: Fontana.

Denison, Edward F. 1967. *Why growth rates differ.* Washington, D.C.: Brookings Institute.

Department of Finance, Canada. 1978. Canada's recent inflation experience.

Despaux, Albert. 1936. *Les dévaluations monétaires dans l'histoire.* Paris: Marcel Rivière.

Devereux, Edward G., Jr. 1950. *Gambling and the social structure: A sociological study of lotteries and horse racing in contemporary America.* Ph.D. dissertation, Harvard University.

Dewey, John. 1930. *Human nature and conduct.* New York: Modern Library.

Downes, D. M., B. P. Davies, M. E. David, and P. Stone. 1976. *Gambling, work and leisure: A study across three areas.* London: Routledge and Kegan.

Drucker, Peter F. 1982. The innovative company. *Wall Street Journal* 26 February.

Duby, Georges. 1973. Les jeunes dans la société aristocratique dans la France du Nord-Ouest an XIIème siècle. In *Hommes et structures du Moyen Age.* Paris: Mouton.

———. 1983. *The Knight the Lady and the Priest: The Making of Modern Marriage in Medieval France.* New York: Pantheon Books.

Duesenberry, James S. 1949. *Income, saving and the theory of human behavior.* Cambridge: Harvard University Press.

Dunlop, John T. 1938. The movement of real and money wage rates. *Economic Journal* 48:413–34.

Durant, Will. 1926. *The story of philosophy.* New York: Washington Square Press, 1961.

Durbin, E.F.M., and John Bowlby. 1939. *Personal aggressiveness and war.* New York: Columbia University Press.

Easterlin, Richard A. 1974. Does economic growth improve the human lot? Some empirical evidence. In *Nations and households in economic growth*, ed. P. David and M. Reder. New York: Academic Press.

Ehrlich, Isaac. 1974. Participation in illegitimate activities: An economic analysis. In *Essays in the economics of crime and punishment.* New York: NBER.

Elias, Norbert. 1973. *The civilizing process.* New York: Urizen Books.

Encyclopedia Britannica. 1976. 15th ed. s.v. English history.

Feldstein, Martin. 1981. Slowing the growth of social security. *Wall Street Journal* 24 September.

Ferro, Marc. 1969. *La grande guerre.* Paris.

Festinger, Leon. 1954. A theory of social comparison processes. *Human Relations* 7:117–40.

Finch, D. 1956. Purchasing power guarantees for deferred payments. *IMF Staff Papers*, 1–22.

Fischer, S. 1975. The demand for index bonds. *Journal of Political Economy* 83:509–34.

———. 1977. Long-term contracts, rational expectations and the optimal money supply rule. *Journal of Political Economy* 85:191–205.

———. 1981. Indexing and inflation. NBER, Working Paper no. 670.

Fisher, Irving. 1911. *The purchasing power of money.* New York.

———. 1928. *The money illusion.* New York: Adelphi.

———. 1932. *Booms and depressions.* New York: Adelphi.

———. 1933a. *Inflation.* New York: Adelphi.

———. 1933b. The debt deflation theory of great depressions. *Econometrica* vol. 1.

———. 1934a. *Mastering the crisis.* London: George Allen.

———. 1934b. *Stable money: A history of the movement.* New York.

Flandrin, Jean-Louis. 1979. *Families in former times: Kinship, household and sexuality.* Cambridge: Cambridge University Press.

Floud, Roderick, and Donald N. McCloskey. 1981. *The economic history of Britain since 1700*. Vol. 1: 1700–1860. Cambridge: Cambridge University Press.

France, Anatole. 1953. *The revolt of the angels*. New York: Heritage Press.

France, Clemens J. 1902. The gambling impulse. *American Journal of Psychology* 13:364–407.

Freud, Sigmund. 1929. A letter to Theodor Reik, April 14, 1929. In Theodor Reik, *Thirty years with Freud*. London: Hogarth Press, 1942, 155–56.

———. 1947. Dostoyevski and patricide. In F. M. Dostoyevski, *Stavrogin's confession*. New York: Lear Publishers.

Friedman, Benjamin M. 1983. Managing the U.S. government deficits in the 1980s. Discussion paper no. 1021, Harvard Institute of Economic Research.

Friedman, Milton. 1962. *Price theory*. Chicago: Aldine.

———. 1974a. Using escalators to help fight inflation. *Fortune* July.

———. 1974b. Monetary correction. In *Essays on inflation and indexation*. Washington, D.C.: American Enterprise Institute.

———. 1984. Why deficits are bad. *Newsweek* 2 January.

———. ed. 1956. *Studies in the quantity theory of money*. Chicago: University of Chicago Press.

———. ed. 1969. *The optimum quantity of money and other essays*. Chicago: Aldine.

———. ed. 1974c. *Indexing and inflation*. AEI Round Table.

Friedman, Milton, and L. J. Savage. 1948. The utility analysis of choices involving risks. *Journal of Political Economy* 56:279–304.

Friedrich, Carl. 1948. *Inevitable peace*. Cambridge: Harvard University Press.

Frisch, Helmut. 1977. Inflation theory 1963–75: A second generation survey. *Journal of Economic Literature* 15:1289–1317.

Fuller, W. A., and G. E. Battese. 1974. Estimation of linear models with gross error structure. *Journal of Econometrics* 2:67–78.

Galbraith, John K. 1958. *The affluent society*. Boston: Houghton Mifflin.

Gambling in America. 1976. Washington, D.C.: Commission on the Review of the National Policy Toward Gambling.

Gambling in Britain. 1972. London: Gallup Social Surveys.

Gardner, Martin. 1981. *Science: Good, bad and bogus*. Buffalo, N.Y.

Gennaro, V. A. 1975. Indexing inflation: Remedy or malady. *Bulletin, Federal Reserve Bank of Philadelphia*, March.

Gigot, Paul A. 1983. The smallest nation has a rare problem. Too much wealth. *Wall Street Journal* 22 September.

Gilpin, Robert. 1981. *War and change in world politics*. Cambridge: Cambridge University Press.

Glasson, Ernest-Désiré. 1882–83. *Histoire du droit et des institutions politiques, civiles et judiciaires de l'Angleterre comparés au droit et aux institutions de la France depuis leurs origines jusqu'à nos jours*. Paris: G. Predone-Lauriel.

Goode, R. 1951. A constant purchasing power savings bond. *National Tax Journal* 4:322–40.

Goody, Jack. 1976. *Production and reproduction: A comparative study of the domestic domain*. Cambridge: Cambridge University Press.

Goody, Jack, Joan Thirsk, and E.P.T. Thompson, eds. 1976. *Family and inheritance: Rural society in Western Europe 1200–1800*. Cambridge: Cambridge University Press.

Gordon, Robert J. 1975. The demand for and supply of inflation. *Journal of Law and Economics* 18:807–36.

Gould, Glenn. 1983. *Glenn Gould: By himself and his friends*. Edited and introduced by John McGreevy. New York: Doubleday.

Grandmont, Jean-Michel, and Guy Laroque. 1976. On temporary Keynesian equilibria. *Revue of Economic Studies* 43:53–67.

Gray, Jo Anna. 1976. Wage indexation: a macroeconomic approach. *Journal of Monetary Economics* 2:221–35.

Greenfield, Sidney M., Arnold Strickon, and Robert T. Aubey. 1979. *Entrepreneurs in cultural context*. Albuquerque: University of New Mexico Press.

Grossman, H. 1975. Is wage response asymmetrical? Paper presented at the Conference on Inflation, Miami.

Gruber, William H., and Raymond Vernon. 1968. The R & D factor in the world trade matrix. In *Technology and competition in international trade*. New York: National Bureau of Economic Research.

Gruber, William H., Dileep Mehta, and Raymond Vernon. 1967. The R & D factor in international trade and international investment of United States industries. *Journal of Political Economy* 75:20–37. In *International Trade and Finance*, ed. R. E. Baldwin and J. D. Richardson. Boston: Little, Brown.

Halevy, Elie. 1930. *The world crisis of 1914–1918: An interpretation*. Oxford: Clarendon Press.

Hall, Stephen S. 1983. Italian-Americans coming into their own. *New York Times Magazine* 15 May.

Hammond, John L., and Barbara Hammond. 1911. *The village labourer 1760–1832*.
———. 1925. *The town labourer 1760–1832*.

Harkabi, Yehoshafat. 1983. *The Bar Kokhba Syndrome*. Chappaqua, N.Y.: Rossel Books.

Harsanyi, John. 1969. Rational choice models of political behavior vs. functionalist and conformist theories. *World Politics* 21:513–38.

Hatcher, John. 1977. *Plague, population and the English economy 1348–1530*. London: Macmillan.

Hayek, Friedrich A. 1973. *Law, Legislation and Liberty*. Chicago: University of Chicago Press.

Heer, Friedrich. *The intellectual history of Europe*, trans. J. Steinberg. New York: World Publishing Co.

Heillener, Karl F. 1967. The population in Europe from the Black Death to the eve of the vital revolution. In *Cambridge economic history of Europe*, vol. 4, ed. E. E. Rich and C. H. Wilson.

Herman, Robert D. 1976 *Gamblers and gambling: Motives, institutions and controls*. Lexington, Mass.: Lexington Books.

Herskovits, Melville J. 1940. *Economic anthropology*. New York: W. W. Norton, 1965.

Hicks, John R. 1937. Mr. Keynes and the "classics": A suggested interpretation. *Econometrica* 5:147–59.
———. 1974. *The crisis in Keynesian economics*. Oxford: Oxford University Press.

Hirshleifer, Jack. 1966. Investment decisions under uncertainty: Applications of the state preference approach. *Quarterly Journal of Economics* 80:252–77.
———. 1977. Economics from a biological viewpoint. *Journal of Law and Economics* 20:1–53.

Hochman, Harold M., and James D. Rogers. 1969. Pareto optimal redistribution. *American Economic Review* 49:542–57.

Holdsworth, Sir William. 1903. *A history of English law*. London: Methivier and Co.

Hollingworth, T. H. 1957. A demographic study of the British ducal families. *Population Studies* 2:4–26.

———. 1964. The demography of the British peerage. Supplement of *Population Studies*. vol. 18.

———. 1969. *Historical demography*, Ithaca: Cornell University Press.

Homans, George Caspar. 1941. *English villagers of the thirteenth century*. New York: Russel & Russel, 1960.

Howard, Michael. 1983. *The causes of wars*. Cambridge: Harvard University Press.

Howell, Cicely. 1976. Peasant inheritance and customs in the Midlands, 1280–1700. In *Family and inheritance*, ed. J. Goody et al. Cambridge: Cambridge University Press.

Hughes, Peter. 1971. Language, history and vision: An approach to 18th century literature. In *The varied pattern: Studies in the 18th century*, ed. P. Hughes and D. Williams. Toronto: A. M. Hakkert.

Huizinga, Johan. 1944. *Homo ludens*. Boston: Beacon Press, 1955.

Jevons, W. S. 1876. *Money and the mechanism of exchange*. New York, 1920.

Johanson, Donald, and Edey Maitland. 1981. *Lucy: The beginnings of humankind*. New York: Warner Books.

Johnson, J. A. 1976. An economic analysis of lotteries. Working Paper nos. 76–16, Department of Economics, McMaster University.

Jones, E. L. 1981. *The European miracle*. London: Cambridge University Press.

Jones, Ernest. 1961. *The life and work of Sigmund Freud*. New York: Basic Books.

Jung, Carl G. 1964. After the catastrophe. In *Civilization in Transition*. Princeton: Princeton University Press, 1978.

Kaplan, H. Roy. 1978. *Lottery winners—How they won and how winning changed their lives*. New York: Harper and Row.

Kendrick, John. 1982. The coming rebound in productivity. *Fortune* 28 June.

Keynes, John M. 1919. *The economic consequences of peace*. London: Macmillan, 1971.

———. 1923. *A tract on monetary reform*. London: Macmillan, 1971.

———. 1930. *A treatise on money*. Vol. 1: *The pure theory of money*. London: Macmillan, 1971.

———. 1931. An economic analysis of unemployment. Lecture 3: The road to recovery. Harris Foundation Lecture, 1931. Reprinted in Keynes, *Collected writings*, ed. Donald Moggridge. London: Macmillan, 1973, 13:358–67.

———. 1936. *The general theory of employment, interest and money*. London: Macmillan.

———. 1939. Relative movements of real wages and output. *Economic Journal* 49:34–51.

———. 1940. How to pay for the war. Reprinted in Keynes, *Collected writings*, ed. Donald Moggridge. London: Macmillan, 9:367–439.

———. 1973a. *The general theory and after. Part 1: Preparation*. Vol. 13 of Keynes, *Collected writings*, ed. Donald Moggridge. London: Macmillan.

———. 1973b. *The general theory and after. Part 2: Defense and Development*. Vol. 14 of Keynes, *Collected writings*, ed. by Donald Moggridge. London: Macmillan.

Kinzer, Stephen. 1983. Brasil fearful of explosion of discontent. *New York Times* 9 October.

Klein, Burton H. 1977. *Dynamic economics*. Cambridge: Harvard University Press.

Knight, Frank H. 1921. *Risk, uncertainty and profit*. New York: Houghton Mifflin.

———. 1947. *Freedom and reform*. New York: Harper & Brothers.

Knox, J. 1964. Index regulated loan contracts. *South African Journal of Economics* 32:237–53.

Kuhn, Thomas S. 1970. *The structure of scientific revolutions*. Chicago: University of Chicago Press.

Kuran, Timur. 1984. Conformity and Conservatism. Working paper, Department of Economics, University of Southern California, November.

Kuznets, Simon. 1953. *Shares of upper income groups in income and savings*. New York: NBER.

Lacroix, Robert, and Claude Montmarquette. 1975. Inflation et indexation: Perspective Canadienne et considérations théoriques. *Canadian Public Policy* 1:185–95.

Lacroix, Robert, and François Dussault. 1981. The spillover effect on public sector wage contracts in Canada. Report no. 8105, Department of Economics, University of Montreal.

Laidler, David, and Michael Parkin. 1975. Inflation—a survey. *Economic Journal* 85:741–97.

Lazear, Edward P., and Robert T. Michael. 1980. Family size and the distribution of real per capita income. *American Economic Review* 70:91–107..

Lazear, Edward P., and Sherwin Rosen. 1981. Rank order tournaments as optimum labor contracts. *Journal of Political Economy* 80:841–64.

Lefèbvre, Charles. 1912. *L'ancien droit des successions*. Vol. 1: Paris: Sirey. (Vol. 2, 1918).

Leibenstein, Harvey. 1957. Economic backwardness and economic growth. New York: John Wiley.

Leibowitz, Nehama. 1980. *Studies in Vayikra (Leviticus)*. Jerusalem: World Zionist Organization.

Leijonhufvud, Axel. 1968. *On Keynesian economics and the economics of Keynes: A study in monetary theory*. New York: Oxford University Press.

Leontief, Wassily W. 1956. Factor proportions and the structure of American trade: Further theoretical and empirical analysis. *Review of Economics and Statistics* 38:386–407.

Lerner, Abba P. 1952. The essential properties of interest and money. *Quarterly Journal of Economics* 66:172–93.

Le Roy Ladurie, Emmanuel. 1976. Family structure and inheritance customs in sixteenth century France. In *Family and inheritance*, ed. J. Goody et al. Cambridge: Cambridge University Press.

Leyser, K. 1968. The German aristocracy from the ninth to the early twelfth century: A historical and cultural sketch. *Past and present* 41:25–53.

Lis, Catharina, and Hugo Soly. 1979. *Poverty and capitalism in pre-industrial Europe*. Atlantic Highlands, N.J.: Humanities Press.

Liviatan, Nissan, and David Levhari. 1977. Risk and the theory of indexed bonds. *American Economic Review* 67:366–76.

Lucas, Robert E. 1973. Some international evidence on output-inflation trade-offs. *American Economic Review* 63:326–39.

Machiavelli, Niccolò. *The Prince*, ed. George Ball. New York: Penguin Books, 1981.

Machlup, Fritz. 1946. Marginal analysis and empirical research. *American Economic Review* 36:519–54.

McCloskey, Donald N. 1981. The Industrial Revolution 1780–1860: A survey. In *The economic history of Britain since 1700*, ed. R. Floud and D. N. McCloskey. Vol. 1: 1700–1860. Cambridge: Cambridge University Press.

McCulloch, John R. 1848. *A treatise on the succession to property vacant by death; Including inquiries on the influence of primogeniture, entails, compulsory partition, foundations etc. over the public interest.* London: Longman, Brown, Green and Longman.

McEvedy, Colin, and Richard Jones. 1978. *Atlas of the world population history.* Middlesex: Penguin.

McFarlane, Alan. 1978. *The origins of English individualism.* Oxford: Basil Blackwell.

Mackinder, Halford. 1949. *Democratic ideals and reality.* New York: W. W. Norton, 1962.

McKinnon, Ian. 1983. What does the public think about deficits? What does Bay Street think about deficits? In *Deficits: How big and how bad?* Toronto: Ontario Economic Council.

McNeill, William H. 1963. *The rise of the West.* Chicago: University of Chicago Press.

———. 1982. *The pursuit of power.* Chicago: University of Chicago Press.

Maine, Henry S. 1885. *Popular government.* Indianapolis: Liberty Classics, 1976.

Maital, Shlomo. 1972. Inflation, taxation and equity. *Economic Journal* 82:158–69.

Mandeville, Bernard. 1932. *The fable of the bees.* Oxford: Clarendon Press, 1957.

March, James G., and Herbert A. Simon. 1958. *Organizations.* New York: John Wiley.

Markowitz, Harry. 1952. The utility of wealth. *Journal of Political Economy* 80:151–58.

Marris, Robin. 1964. *The economic theory of "managerial" capitalism.* Glencoe, Ill: Free Press.

Marshall, A. 1887. Remedies for fluctuations of general prices. In *Memorials of Alfred Marshall*, 1925.

———. 1925. *Principles of economics.* 8th ed. London: Macmillan.

Martin, Xavier. 1972. *Le principe d'egalité dans les successions roturières en Anjou et dans le Maine.* Paris: Presses Universitaires de France.

May, Mark. 1943. *A social psychology of war and peace.* New Haven: Yale University Press.

Mead, Margaret. 1940. Warfare is only an invention—not a biological necessity. *Asia* 40:402–5.

Merton, Robert K. 1957. *Social theory and social structure.* New York: Free Press.

Mill, John S. 1848. *Principles of political economy.* London: Routledge & Sons, Limited [n.d.].

Modelski, George. 1978. The long cycle of global politics and the nation-state. *Comparative Studies in Society and History.* 20:214–35.

Modigliani, Franco. 1983. Government deficits, inflation, and future generations. In *Deficits: How big and how bad?*, ed D. W. Conklin and T. J. Courchene. Toronto: Ontario Economic Council.

Mokyr, J. 1977. Demand vs. supply in the Industrial Revolution. *Journal of Economic History* 37:981–1008.

———. 1984. *The economics of the Industrial Revolution*. Totowa, N.J.: Roman and Allenheld.

Moore, Wilbert E. 1968. Social change. In *International Encyclopedia of the Social Sciences*, ed. David Sills. New York: Crowell, Collier, and Macmillan, 14:365–75.

Morgenthau, Hans. 1960. *Politics among nations*. 3d ed. New York: Alfred A. Knopf.

Morison, Samuel E. 1965. *The Oxford history of the American people*. New York: New American Library, 1972.

Mueller, Gerhard O. W. 1983. Grimm "fairy" tales as the lore of law. *New York Times* 9 October.

Musgrave, Richard A., and Peggy B. Musgrave. 1973. *Public finance in theory and practice*. New York: McGraw-Hill.

Nakamura, J. I. 1981. Human capital accumulation in premodern rural Japan. *Journal of Economic History* 41:263–81.

Nelson, Benjamin. 1969. *The idea of usury: From tribal brotherhood to universal otherhood*. 2d ed. Chicago: University of Chicago Press.

Neurisse, André. 1978. *Histoire de l'impôt*. Paris.

Niebuhr, Reinhold. 1938. *Beyond tragedy*. New York: Charles Scribner's Sons.

Nisbet, Robert A. 1969. *Social change and history*. Oxford: Oxford University Press.

———. 1973. *The social philosophers*. New York: Washington Square Books, 1982.

———. 1982. *Prejudices*. Cambridge: Harvard University Press.

North, Douglass C. 1981. *Structure and change in economic history*. New York: W. W. Norton.

North, Douglass C., and Robert P. Thomas. 1973. *The rise of the Western world: A new economic history*. Cambridge: Cambridge University Press.

Olson, Mancur. 1978. The political economy of comparative growth rates. Working Paper, Department of Economics, University of Maryland.

———. 1980. Stagflation and the political economy of the decline in productivity. *American Economic Review* Papers and Proceedings: 143–49.

———. 1982. *The rise and decline of nations*. New Haven: Yale University Press.

Ohmae, Kenichi. 1982a. *The mind of the strategist*, New York: McGraw-Hill.

———. 1982b. Japan's entrepreneurs battle the goliaths. *Wall Street Journal* 20 December.

Opie, Iona, and Peter Opie. 1974. *The classic fairy tales*. Oxford: Oxford University Press.

Organski, A.F.K. 1968. *World politics*. 2d ed. New York: A. A. Knopf.

Organski, A.F.K., and Jacek Kugler. 1980. *The war ledger*. Chicago: University of Chicago Press.

Orwell, George. 1945. Notes on nationalism. In *Decline of the English murder and other essays*. New York: Penguin Books, 1980.

Parks, Richard W. 1975. Inflation and relative price variability. *Journal of Political Economy* 83:241–54.

Paul, F., and D. Williams, eds. 1972. *The triumph of culture: 18th century perspectives*. Toronto: A. M. Hakkert.

Peters, Thomas J., and Robert H. Waterman, Jr. 1982. *In search of excellence*. New York: Harper and Row.

Phelps, Edmund S., ed. 1970. *Microeconomic foundations of employment and inflation theory*. New York: W. W. Norton.

Pickering, George. 1974. *Creative malady* New York: Delta Books.

Pike, Luke Owen. 1873. *A history of crime in England*. Montclair, N.J.: Patterson Smith.

Pinson, Koppel S. 1954. *Modern Germany* New York: Macmillan.

Plato. *Timaeus and Critias*. Baltimore: Penguin Books. 1974.

———. *Republic*, ed. S. Buchanan. Middlesex: Penguin Books, 1978.

Plumb, J. H. 1972. The public, literature and the arts in the 18th century. In *The triumph of culture: 18th century perspectives*, ed. Paul Fritz and David Williams. Toronto: A. M. Hakkert.

Pollock, Frederick, and F. W. Maitland. 1898. *The history of English law*. Cambridge: Cambridge University Press, 1968.

Radzinowicz, Leon. 1956. *A history of English criminal law and its administration from 1750*, vol. 2. London: Stevens & Sons.

Rapoport, Anatol. 1982. Introduction to von Clausewitz, *On War*. London: Penguin Books.

Report of the Royal Commission on Betting, Lotteries and Gaming. 1951. London: Royal Commission, HMSO.

Rimlinger, Gaston V. 1971. *Welfare policy and industrialization in Europe, America and Russia*. New York: John Wiley.

———. 1982. The historical analysis of national welfare systems. In *Explorations in new economic history: Essays in honor of Douglass C. North*, ed. Roger L. Ransom, Richard Sutch, and Gary M. Walton. New York: Academic Press.

Robert Sylvestre Marketing Ltd. 1977. *Le marché Québécois des loteries*. Montreal.

Robbins, Helen. 1928. Black death in France and England. *Journal of Political Economy* 36:447–79.

Robinson, Joan. 1962. *Economic philosophy*. London: Penguin Books.

Rosenblum, Roger. 1971. The dawn of British romantic painting, 1760–1780. In *The varied pattern: Studies in the 18th century*, ed. P. Hughes and D. Williams. Toronto: A. M. Hakkert.

Rudé, G. 1972. Popular protest in 18th century Europe. In *The triumph of culture: 18th century perspectives*, ed. F. Paul and D. Williams. Toronto: A. M. Hakkert.

Russel, J. G. 1972. Population in Europe 500–1500. In *The Fontana economic history of Europe*, ed. C. M. Cipolla. Vol. 1: *The Middle Ages*. London: Fontana Books.

Sabean, David. 1976. Aspects of kinship behavior and property in rural Western Europe before 1800. In *Family and inheritance*, ed. J. Goody et al. Cambridge: Cambridge University Press.

Sadka, Efraim. 1976. On progressive income taxation. *American Economic Review* 66:931–35.

Schelling, Thomas C. 1958. The strategy of conflict. Prospectus for a reorientation of game theory. *Journal of Conflict Resolution* 2:203–64.

Scherer, F. M. 1980. *Industrial market structure and economic performance*. Chicago: Rand McNally.

Schultz, T. W. 1961. Investment in human capital. *American Economic Review* 51:1–17.

Schumpeter, Joseph A. 1919, 1927. *Imperialism and social classes*, ed. Paul Sweezy. New York: Augustus M. Kelley, 1951.

———. 1942. *Capitalism, socialism and democracy*. New York: Harper & Row.

———. 1954. *History of economic analysis*. London.

Schaffer, Peter. 1983. Paying homage to Mozart. *New York Times Magazine* 2 September.

Sheehan, Michael M. 1963. *The will in Medieval England from the conversion of the Anglo-Saxons to the end of the thirteenth century*. Toronto: Pontificat Institute of Medieval Studies.

Shils, Edward. 1981. *Tradition*. Chicago: University of Chicago Press.

Simon, Herbert A. 1957. *Administrative Behaviour*. New York: Macmillan.

———. 1959. Theories of decision-making in economics and behavioral science. In *Microeconomics: Selected readings*. New York: W. W. Norton, 1971.

Smith, Adam. The history of astronomy. In *Essays on philosophical subjects*. Indianapolis: Liberty Classics, 1980.

Smith, William Carlson. 1953. *The Stepchild*. Chicago: University of Chicago Press.

Solman, Paul, and Thomas Friedman. 1982. *Life and death on the corporate battlefield*. New York: Simon and Schuster.

Solow, Robert M. 1957. Technical Change and the Aggregate Production Function. *Review of Economics and Statistics* 39.

Stern, F. 1977. *Gold and iron: Bismarck, Bleichroder, and the building of the German empire*. London: George Allen.

Stevens, Neil A. 1981. Indexation of social security benefits—A reform in need of reform. *Federal Reserve Bank of St. Louis* 63:1–70.

Stigler, George J. 1975. *The citizen and the state*. Chicago: University of Chicago Press.

Stoessinger, John G. 1978. *Why nations go to war*. 2d ed. New York: St. Martin's Press.

Stone, Lawrence. 1979. *The family, sex and marriage in England 1500–1800*. Middlesex: Penguin.

Sullivan, George. 1972. *By chance a winner*. New York: Dodd, Mead and Co.

Tarshis, Lorie. 1939. Changes in real and money wages. *Economic Journal* 49:150–54.

Taylor, Gordon R. 1979. *The natural history of the mind*. New York: Penguin Books, 1981.

Tec, Nechama. 1964. *Gambling in Sweden*. Totowa, N.J.: Bedminster Press.

Thirsk, Joan. 1976. The European debate on customs of inheritance 1500–1700. In *Family and inheritance*, ed. J. Goody et al. Cambridge: Cambridge University Press.

Thomas, Lewis. 1974. *The lives of a cell*. New York: Viking Press.

Thompson, E. P. 1975. The crime of anonymity. In *Albion's fatal tree*. ed. D. Hay et al. New York: Pantheon Books.

Thucydides. *History of the Peloponnesian War*, trans. Rex Warner. New York: Penguin, 1954.

Thurow, Lester C. 1971. The income distribution as a pure public good. *Quarterly Journal of Economics* 85:327–36.

238   References

Todd, Emmanuel. 1979. *Le fou et le prolétaire*. Paris.

Totman, Richard. 1979. *Social Causes of Illness*. New York: Pantheon Books.

Toynbee, Arnold. 1884. *The Industrial Revolution*. Boston: Beacon Press, 1956.

Toynbee, Arnold, Jr. 1966. *Change and habit: The challenge of our time*. London: Oxford University Press.

Trevithick, James A. 1975. Keynes, inflation and money illusion. *Economic Journal* 85:101–13.

Tuchman, Barbara W. 1981. *Practicing history*. New York: Ballantine Books.

U.S. Bureau of the Census. 1975. *Historical statistics of the U.S. from colonial times to 1970*. Washington D.C.: Government Printing Office.

———. 1981. *Statistical abstract of the U.S., 1980*. Washington, D.C.: Government Printing Office.

U.S. Department of Commerce. 1977. *Technology assessment and forecast*. 7th report. Washington, D.C.: Government Printing Office.

Veblen, Thornstein. 1933. *The engineers and the price system*. New York: Viking Press.

Vernon, Raymond. 1966. International investment and international trade in the product cycle. *Quarterly Journal of Economics* 80:190–207.

Vilar, Pierre. 1966. Problems of the formation of capitalism. In *The rise of capitalism*, ed. D. Landes. New York: Macmillan.

Waelder, Robert. 1967. *Progress and revolution*. New York: International Universities Press.

Wallace, William H., and William E. Cullison. 1979. *Measuring price changes: A study of the price indexes*. Federal Reserve Bank of Richmond.

Waltz, Kenneth N. 1954. *Man, the state and war*. New York: Columbia University Press.

Wedgwood, Josiah. 1929. *The economics of inheritance*. London: Routledge and Sons.

Weinstein, David, and Lilian Deitch. 1974. *The impact of legalized gambling: The socioeconomic consequences of lotteries and off-track betting*. New York: Praeger.

Weisser, Michael A. 1979. *Crime and punishment in early modern Europe*. Atlantic Highlands, N.J.: Humanities Press.

Westrup, G. W. 1936. *Family property and patrias potestas*. Oxford: Humphrey Milford at the Oxford University Press.

Wilde, Oscar. 1966. The soul of man under socialism. In *The complete works of Oscar Wilde*. London: Collins.

Williamson, Jeffrey G., and Peter H. Lindert. 1980. *American inequality: A macroeconomic history*. New York: Academic Press.

Williamson, Oliver E. 1964. *The economics of discretionary behavior: Managerial objectives in a theory of the firm*. Englewood Cliffs, N.J.: Prentice-Hall.

Wilson, Edmund. 1983. The partnership of Marx and Engels. In *The portable Edmund Wilson*. ed. Lewis M. Daleney. New York: Penguin Books. 253–301.

Wilton, David A. 1980. An analysis of Canadian wage contracts with cost-of-living allowance clauses. Document no. 165, Economic Council of Canada, Ottawa.

Wright, Quincy. *A Study of War*. Chicago: University of Chicago Press, 1964.

Wrigley, E. A., and R. S. Schoffield. 1981. *The population history of England.* Cambridge: Cambridge University Press.

Yver, Jean. 1966. *Egalité entre héritiers et exclusions des enfants dotés: Essai de géographie coutumière.* Paris: Editions Sirey.

———. 1984. Every man's budget. *Wall Street Journal* 2 February.

Index

France, 12, 14, 23, 133, 135, 176, 179, 192, 193, 207n.43, 212nn. 8, 18, 213n.19; inheritance laws in, 117–29
France, A., 36, 196, 197
France, C., 66
Freud, S., 63, 100, 205nn. 18, 19, 20
Friedman, B. M., 218n.39
Friedman, M., 63, 71, 96–97, 137, 184, 203n.34, 213n.5, 217n.38, 218n.39, 220nn. 29, 30
Friedman, T., 99, 209n.62
Friedrich, C., 28
Frisch, H., 219n.13
Fuller, W. A., 88

Galbraith, J. K., 41
Gamblers: age of, 58–61; income of, 57–62; leisure activities of, 63, 65; social class of, 61, 104–5. See also Gambling; Lotteries
Gambling: and crime, 64–66; during the Industrial Revolution, 104–6; in Canada 58–62; in England, 64–65; in Sweden, 61–62; in the U.S., 61, 63–66; opinions on, 63–66; theory of, 38–43
Games of chance, 4, 57; and religious beliefs, 30–31. See also Gambling; Lotteries
Game theory, 202n.31
Gardner, M., 220n.2
Gennaro, V. A., 220n.28
Germany, 12–13, 20, 24, 121, 124, 133, 192, 193, 199n.13, 200n.14, 201nn. 17, 24, 207n.43. See also Bismarck, O. von
Gibbon, E., 187
Gigot, P. A., 23
Gilpin, R., 186, 187, 188, 193, 200n.14, 201n.18
Glasson, E. D., 212n.16
Goode, R., 220n.28
Goody, J., 117, 119, 120, 121
Gordon, R. J., 213n.3, 219n.13
Gould, G., 196
Government: ideal, 18; expenditures, 153–57. See also Customs; Deficits; Inflation; Population; Religion
Grandmont, J. M., 144
Gray, J. A., 218n.7
Greenfield, S. M., 212n.11
Greiff, B., 207n.52
Grossman, H., 220n.30
Growth, 199nn. 3, 14, 202n.27
Gruber, W. H., 91, 95, 96

Halevy, E., 199n.13
Hall, S. S., 208n.56
Hammond, B., 210n.5

Hammond, J. L., 210n.5
Harbaki, Y., 31, 201n.17
Harsanyi, J., 187
Harrod, R. F., 140, 214nn. 10, 15
Hatcher, J., 125
Hayek, F. A. 213n.20
Heer, F., 101
Heillener, K. F., 213n.19
Herman, R. D., 63
Herskovits, M. J., 22
Hicks, J. R., 144, 216nn. 25, 26, 27, 220n.30
Hierarchy, 79. See also Classes, social; Disorder; Leapfrogging; "Outdone by fellows"; Wealth distribution; Wealth redistribution
Hirshleifer, J., 203nn. 33, 34
Hobbes, T., 19, 55
Hochman, H. M., 71, 72
Holdsworth, W., 123, 124, 128, 130, 201n.23, 212nn. 14, 16
Hollingsworth, T. H., 125
Homans, G. C., 211n.7
Howard, M., 14, 20
Hughes, P., 114, 211nn. 11, 12
Huizinga, J., 30, 31, 202n.29

Ideas: betting on, 4, 29–37, 45–49, 204nn. 38, 42, 210n.1, 217n.34, 220n.2; political, 6–7, 9, 47–48, 53–56; revolutionary, 9, 68–69, 72–74; 200n.14; and social order 18, 186–90. See also Art; Creativity; Entrepreneurship; Productivity; Thinking
Iks, the, 195
Indexation: and money supply, 166–69, 182; and monetarist framework, 184–85; and Keynesian framework 184–85; in Canada, 169–83; in the public sector, 172–73; of taxes, 173; of transfer payments, 173; of wages, 170–73, 183. See also Index, numbers
Index: consumer price (CPI), 169, 173–76, 219n.26, 220n.27; numbers, 166–68, 217n.37. See also Indexation
Industrial Revolution: arts during, 114–15; creativity during, 112–16, 210n.9; gambling during, 104–6; opinions on, 102–4, 112–13; social instability during, 106–12
Inflation, 139, 143–44, 158–63, 169, 181, 215n.23, 216n.33; after wars, 131–36, 162–63; "cost-push," 134, 213n.3, 219n.13; hyperinflation, 137; Keynes's definition of, 142. See also Debt; Deficits; Indexation; Index numbers; Money supply